MICROWAVE REMOTE SENSING
of the
EARTH SYSTEM

Studies in
Geophysical Optics and Remote Sensing
Series Editor: Adarsh Deepak
Published Volumes and Volumes in Preparation

Selected papers from the

Symposium on Microwave Remote Sensing, which was one of the Symposia organized by the International Association of Meteorology and Atmospheric Physics (IAMAP), at the **XIX General Assembly of the International Union of Geodesy and Geophysics.** The General Assembly was held at the University of British Columbia, Vancouver, Canada, 9–22 August 1987. The Symposium was convened by Professor Alain Chedin. Preparations for the meeting were made by the IUGG Local Organizing Committee.

MICROWAVE REMOTE SENSING of the EARTH SYSTEM

Edited by

Alain Chedin

Laboratoire de Métérologie Dynamique du
Centre National de la Recherche Scientifique

A. DEEPAK Publishing   1989
A Division of Science and Technology Corporation
Hampton, Virginia USA

A. DEEPAK Publishing
A Division of Science and Technology Corporation
101 Research Drive
Hampton, Virginia 23666-1340   USA

Library of Congress Cataloging-in-Publication Data

Microwave remote sensing of the earth system.
(Studies in geophysical optics and remote sensing)
Papers given at the Microwave Remote Sensing Symposium
(IAMAP) of the 19th IUGG General Assembly held in
Vancouver, Canada, August 9–22, 1987
Bibliography: p.
Includes index.
1. Remote Sensing—Congresses. 2. Microwave devices—
Congresses. I. Chedin, Alain. II. Microwave Remote
Sensing Symposium (1987 : Vancouver, Canada) III. Series.
G70.4.M5   1989   551'.028   89–1335
ISBN 0–937194–17–4

CONTENTS

PREFACE

Observation of the Earth's system in the microwaves is a domain of research of rapidly growing interest. This is due to the almost complete insensitivity to clouds of microwave radiation, thus allowing all weather measurements on the one hand, and to the existence of signatures, particular to these wavelengths, like those of sea ice, cloud liquid water, trace gases, ice crystals, soil moisture, precipitations, etc., on the other. The book reports on recent theoretical and experimental developments in this field of research, carried out by a variety of expert groups.

The volume is a collection of papers that are extended versions of the papers given at the Microwave Remote Sensing Symposium (IAMAP) of the XIX IUGG General Assembly, held in Vancouver, Canada, August 9–22, 1987. It includes papers covering a relatively wide range of applications. Three papers deal with the theory of radiative transfer, with an intercomparison of three important forward models, and with applications to two major future space experiments—the Microwave Limb Sounder on UARS (Upper Air Research Satellite) for trace gases monitoring in the middle atmosphere and the Advanced Microwave Sounding Unit on the TIROS-NEXT US operational weather satellites.

Retrieval algorithms may benefit greatly from an a priori knowledge of the medium observed (shape of the temperature or water vapor vertical profiles—two papers) or of the instrument characteristics that may be complex in the microwaves (the non-uniformity of the fields of view—one paper). The other five papers report on study cases, involving a variety of instruments such as passive multichannel radiometers for imaging or sounding or active doppler radars, for example, and cover complex atmospheric or surface phenomena.

Passive radiometry is applied to the retrieval of a number of parameters such as atmospheric water vapor or liquid water, atmospheric temperature, soil moisture, oil slick, and ice types. The frequency domain covered ranges from 9 GHz to 183 GHz and observations are made from the ground, ship or space. These results should be of great help in processing data acquired by past, present, or planned microwave spaceborne instruments on NIMBUS-7, DMSP, TOPEX-POSEIDON, TIROS-NEXT, METEOSAT 2nd Generation, and so forth.

Active radiometry is illustrated by two papers. The first one attacks the important and difficult problem of the determination of the three-dimensional wind velocity structure and presents an application to the study of the horizontal velocity field in the marine atmospheric planetary boundary layer. The second deals with tall convective storms through the examination of radar echoes.

It is hoped that the papers included in this volume will make it an important source of information for researchers already in or intending to enter the very active, promising, and powerful domain of microwave remote sensing.

The editor wishes to acknowledge the cooperation of session chairmen, speakers, and participants for making this a stimulating and valuable symposium for everyone.

A. CHEDIN

ATMOSPHERIC TRANSMITTANCE MODELING IN THE SUBMILLIMETER REGION

Lewis L. Smith

Grumman Corporate Research Center
Bethpage, New York 11714, USA

ABSTRACT

Three leading line-by-line submillimeter models (LINETRAN, FASCODE, and MPM) are compared with existing data. All three models compare favorably with both horizontal and vertical atmospheric path measurements.

1. INTRODUCTION

The submillimeter region of the spectrum is of interest to, e.g., astronomers studying molecules in space, spectroscopists investigating rotational transitions of molecules, and scientists studying the propagation properties of the atmosphere. Three line-by-line submillimeter models: LINETRAN (Smith and Hilgeman, 1981); Fast Atmospheric Signature Code - FASCODE (Clough et al., 1981); and Millimeter-wave Propagation Model - MPM (Liebe, 1983) have been developed to predict atmospheric transmittances and radiances. In this paper, these three models are compared and validated with horizontal and vertical atmospheric path data.

2. MODEL DESCRIPTIONS

All three line-by-line submillimeter models -- LINETRAN, FASCODE, and MPM -- predict the atmospheric spectral transmission for a multilayer atmosphere.

The Grumman-developed LINETRAN model and the AFGL FASCODE model both use the AFGL HITRAN line parameter atlas, the Van Vleck-Huber line profile, and the H_2O continuum formulation from FASCODE. The Liebe MPM model uses a selected set of H_2O and O_2 lines, the Van Vleck-Weisskopf line profile, and a Liebe H_2O continuum. The advantage of LINETRAN over FASCODE is that it is easier to implement. In addition, it accesses a much larger line data base than the data base accessed by MPM.

MICROWAVE REMOTE SENSING
of the EARTH SYSTEM
Alain Chedin (Ed.)

1

The contribution to the optical depth due to the molecular lines at a frequency ν is given by

$$k_1(\nu) = \sum_j \sum_i S_i(T_j)\, n_{ij}\, \phi_i(\nu, T_j, P_j) \tag{1}$$

where

T_j is the temperature of the j-th atmospheric layer

$S_i^j(T_j)$ is the line strength for the i-th molecule in the j-th layer

n_{ij} is the column density of the i-th molecule in the j-th layer

and

$\phi_i(\nu, T_j, P_j)$ is the Voigt line profile for the i-th molecule in the j-th layer.

The Voigt profile is the convolution of the collision and Doppler profiles and is a function of the temperature T_j and the pressure P_j of the j-th layer as well as the line parameters of the i-th molecule.

TABLE 1. COLLISION LINE PROFILES*

1. Simple Lorentz $\dfrac{\alpha_i}{\pi} \dfrac{1}{(\nu-\nu_o)^2 + \alpha_i^2}$

2. Gross $\dfrac{\alpha_i}{\pi}\left(\dfrac{\nu_o}{\nu}\right) \dfrac{4\nu^2}{(\nu^2 - \nu_o^2)^2 + 4\,\alpha_i^2\,\nu^2}$

3. Van Vleck-Weisskopf

$$\dfrac{\alpha_i}{\pi}\left(\dfrac{\nu}{\nu_o}\right)\left[\dfrac{1}{(\nu-\nu_o)^2 + \alpha_i^2} + \dfrac{1}{(\nu+\nu_o)^2 + \alpha_i^2}\right]$$

4. Modified Van Vleck-Weisskopf

$$\dfrac{\alpha_i}{\pi}\left(\dfrac{\nu}{\nu_o}\right)\dfrac{1-e^{-h\nu/kT}}{1-e^{-h\nu_o/kT}}\left[\dfrac{1}{(\nu-\nu_o)^2 + \alpha_i^2} + \dfrac{1}{(\nu+\nu_o)^2 + \alpha_i^2}\right]$$

5. Van Vleck-Huber

$$\dfrac{\alpha_i}{\pi}\left(\dfrac{\nu}{\nu_o}\right)\dfrac{\tanh\left(\dfrac{h\nu}{2kT}\right)}{\tanh\left(\dfrac{h\nu_o}{2kT}\right)}\left[\dfrac{1}{(\nu-\nu_o)^2 + \alpha_i^2} + \dfrac{1}{(\nu+\nu_o)^2 + \alpha_i^2}\right]$$

* α_i is collisional half-width at 1013 mb and 296 K
  ν_o is the transition frequency
  h is Planck's constant
  k is Boltzmann's constant

Both LINETRAN and FASCODE use the Van Vleck-Huber collision line profile, while MPM uses the Van Vleck-Weisskopf profile, modified by Rosenkranz (1975). Table 1 lists these profiles along with three others modified for the submillimeter region according to Falcone (1969). All five profiles are compared in Fig. 1 for a line with transition frequency ν_o = 50 cm^{-1} and collisional halfwidth α_i = 0.1 cm^{-1}. At the line center all the profiles approach the same value, $(\pi\alpha_i)^{-1}$, and diverge when $|\nu-\nu_o| > 2$ cm^{-1}. Only the simple Lorentz is symmetric about ν_o, with the Van Vleck-Huber profile being the most asymmetric.

The Van Vleck-Huber and the Van Vleck-Weisskopf profiles differ appreciably in the far wings, and this is taken into account by the different forms of the water continuum in the models. FASCODE and LINETRAN calculate the line profile out to $|\nu - \nu_o| = 25$ cm^{-1}. The contribution to the optical depth due to the water continuum, discussed by Clough et al. (1981), is given by

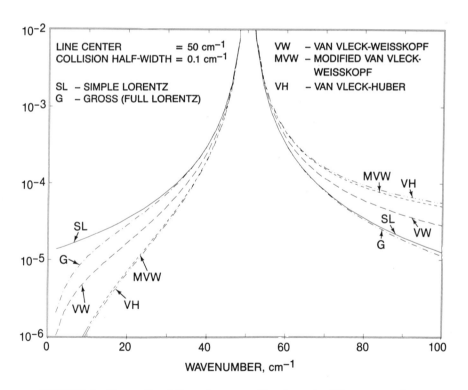

FIGURE 1. Normalized line shape profiles in submillimeter region. The models have been computed for a line centered at 50 cm^{-1} and collision half-width of 0.1 cm^{-1}.

$$k_c(v) = v \sum_j n_{sj}\tanh(hv/hT) \left[\left(\frac{n_{sj}}{n_{oj}}\right) \tilde{C}_s(v,T_j) + \left(\frac{n_{fj}}{n_{oj}}\right) \tilde{C}_f(v,T_j)\right] \qquad (2)$$

where

n_{sj} is the column density of water in the j-th layer

n_{fj} is the column density of all other molecules
              in the j-th layer

and

n_{oj} is the total column density defined at 1013 mb
              and 296 K.

Values of $\tilde{C}_s(v,T_j)$ and $\tilde{C}_f(v,T_j)$ can be derived from the tabu-
lations given in FASCODE.

The MPM program uses a water continuum formulation described by
Liebe (1985, 1987) to bring experimental and predicted attenuation
data into agreement at 4.6 cm^{-1} (138 GHz).

LINETRAN and FASCODE use the 1986 version of the AFGL HITRAN line
parameter atlas (Rothman et al., 1987) which contains information
on 28 molecular species between 0 and 17,900 cm^{-1}. The MPM model uses
48 O_2 lines and 30 H_2O lines between 0 and 33 cm^{-1} (Liebe, 1983).

3. MODEL COMPARISONS WITH DATA

The atmospheric transmission measurements of Rice and Ade (1979)
observed from the ground looking up are compared to the corresponding
LINETRAN prediction (Fig. 2) for a line-of-sight water vapor
concentration of 9.3 ± 1.0 mm. The three-layer atmospheric input
parameters (P,T, and molecular column densities) are specified by Rice
and Ade, and the calculations have been degraded to 0.12 cm^{-1}
resolution. The strong lines of H_2O and O_2 are seen as well as the
weaker lines of O_3 in the region 7-10 cm^{-1}. The LINETRAN calculations
fall a little below the data for $v < 6$ cm^{-1} and a little above for $v >$
6 cm^{-1}. Rice and Ade do not show any error bars for their
measurements, hence the accuracy of the data is somewhat uncertain.
Considering the coarseness of a three-layer atmospheric model, the
uncertainty in the water vapor determination at the time of the
measurements, as well as the uncertainty in the data accuracy, the
LINETRAN results predict the submillimeter transmission of the
atmosphere quite accurately.

It should be noted that Clough et al. (1981) fit a two-layer
FASCOD1B calculation to the Rice and Ade data with slightly better
results than are shown in Fig. 2. This is hard to explain as LINETRAN
and FASCOD1C models give essentially the same predictions for the Rice
and Ade case, as shown in Fig. 3 where the resolution has been
improved to 0.01 cm^{-1}. The weak O_3 lines in the region 7-10 cm^{-1} are
now quite prominent compared to the calculations of Fig. 2. Figure 3

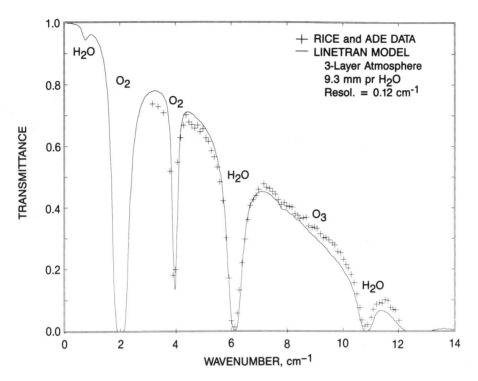

FIGURE 2. LINETRAN transmittance compared to Rice and Ade data. Three-layer atmospheric model, 9.3 mm H_2O, resolution 0.12 cm^{-1}.

shows that LINETRAN and FASCOD1C predict similar transmittances, as might be expected, since they both use the Van Vleck-Huber collision line profile, the FASCODE H_2O continuum, and the HITRAN line parameter atlas. One advantage LINETRAN has over FASCODE is that it is easier to implement. While it takes longer to run a given calculation with LINETRAN than FASCODE, this is more than outweighed by the convenience of having fewer than 500 lines of LINETRAN code, compared to 8400 lines for FASCOD1C and 23,000 lines for FASCOD2.

LINETRAN, FASCODE, and MPM model predictions are compared to horizontal path attenuation measurements of Furashov et al. (1984) in Figs. 4 and 5. The attenuation γ (dB/km) is related to the transmittance τ by γ = -10 log$_{10}$ (τ). The water vapor data in Fig. 4 were taken using a multipass vacuum cell with an optical path of 140 m (P = 973 mb, T = 298.7 K, H_2O = 19 g/m^3). The MPM model calculations were taken from Liebe (1987). The measurements fall between MPM and LINETRAN/FASCODE in the 11.0 - 12.2 cm^{-1} window region, while MPM is closer to the data in the 13.0 - 14.3 cm^{-1} window. The small differences in the atmospheric windows between MPM and LINETRAN/FASCODE are to be expected due to the different collision line profiles and water continuum formulations used.

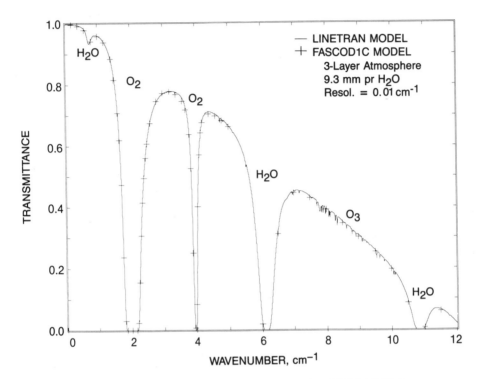

FIGURE 3. LINETRAN transmittance compared to FASCOD1C for Rice and Ade case. Three-layer atmospheric model, 9.3 mm H_2O, resolution 0.01 cm^{-1}.

Horizontal measurements of Furashov et al. (1984) over a 1.58 km path are shown in Fig. 5 (P = 969 mb, T = 281.7 K, H_2O = 8.5 g/m^3). All three models agree closely with the measurements and predict essentially the same attenuation in the window regions. The MPM prediction starts to diverge from LINETRAN/FASCODE in the window regions as the water vapor density increases.

The atmospheric transmission in the submillimeter region for obervations at the zenith for two altitudes, 1.5 km and 4.0 km, is shown in Figs. 6 and 7. The calculations used the U.S. Standard Atmosphere (McClatchey et al., 1972), and the resolution was 0.002 cm^{-1}. For the 1.5-km case (Fig. 6) the atmosphere is opaque for $\nu > 14$ cm^{-1}, while for 4.0 km (Fig. 7) the transmittance increases to 0.17 in the centers of the 22.2 cm^{-1} (450 µm) and 28.5 cm^{-1} (350 µm) regions. This results from the lower precipitable water vapor and atmospheric pressures at 4.0 km. Astronomers have successfully observed in these two submillimeter windows on Mauna Kea, Hawaii (Whitcomb et al., 1979; Roellig et al., 1986).

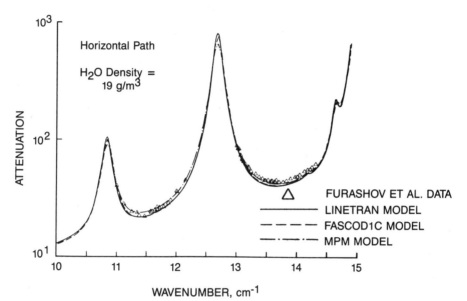

FIGURE 4. Attenuation in dB/km for LINETRAN, FASCOD1C, and MPM models compared to horizontal path measurements of Furashov et al. (H_2O density = 19 g/m^3).

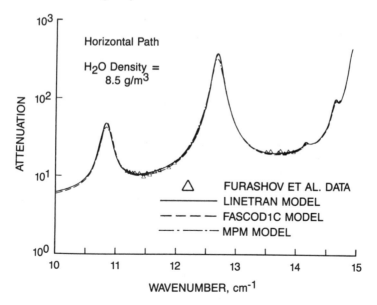

FIGURE 5. Attenuation in dB/km for LINETRAN, FASCOD1C, and MPM models compared to horizontal path measurements of Furashov et al. (H_2O density = 8.5 g/m^3).

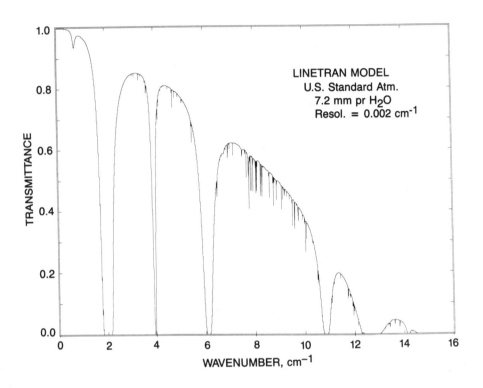

FIGURE 6. Atmospheric transmittance: 1.5 km altitude to space, 0-deg zenith angle, U.S. Standard Atmosphere.

4. SUMMARY

In summary, three line-by-line submillimeter models, LINETRAN, FASCODE, and MPM are compared to one another and to horizontal and vertical atmospheric transmission data. LINETRAN and FASCODE give essentially identical results. MPM differs slightly, but all three models compare favorably with the measurements.

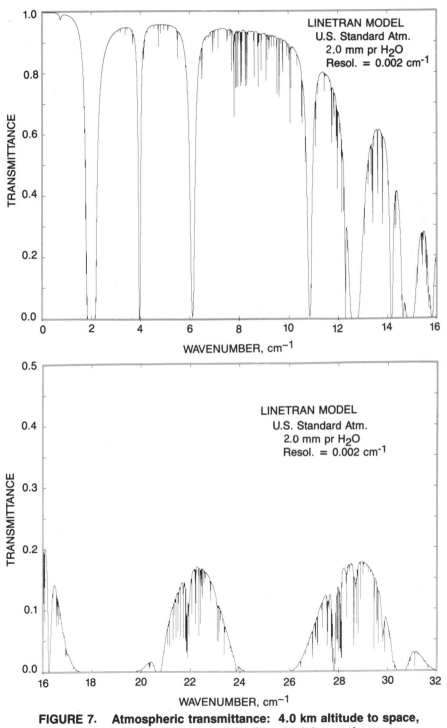

FIGURE 7. Atmospheric transmittance: 4.0 km altitude to space, 0-deg zenith angle, U.S. Standard Atmosphere.

REFERENCES

Clough, S.A., F.X. Kneizys, L.S. Rothman, and W.O. Gallery, 1981: Atmospheric Spectral Trasmittance and Radiance: FASCOD1B, in Atmospheric Transmission (R.W. Fenn, ed.), Proc. SPIE 277, 152, Intl. Soc. for Optical Engineering, P.O. Box 10, Bellingham, WA 98227.

Falcone, V.J., Jr., 1969: A Persistent Error in Collision-Broadened Line Shape Factors, Appl. Opt., 8, 2362.

Furashov, N.I., Yu. V. Katkov, and V. Ya. Ryadov, 1984: On the Anomalies in Submillimeter Absorption Spectrum of Atmospheric Water Vapor, Intl. J. of Infrared and Millimeter Waves, 5, 971.

Liebe, H.J., 1983: An Atmospheric Millimeter Wave Propagation Model, NTIA Report 83-137, Nat. Telecomm. and Inform. Admin., Boulder, CO 80303.

Liebe, H.J., 1985: An Updated Model for Millimeter Wave Propagation in Moist Air, Radio Sci., 20, 1069.

Liebe, H.J., 1987: A Contribution to Modeling Atmospheric Millimeter-wave Properties, Frequenz, 41, 31.

McClatchey, R.A., R.W. Fenn, J.E.A. Selby, F.E. Volz, and J.S. Garing, 1972: Optical Properties of the Atmosphere (Third Edition), AFCRL-72-0497, Air Force Geophy. Labs., Hanscom AFB, MA 01731.

Rice, D.P., and P.A.R. Ade, 1979: Absolute Measurements of the Atmospheric Transparency at Short Millimetre Wavelengths, Infrared Phys., 19, 575.

Roellig, T.L., E.E. Becklin, C.D. Impey, and M.W. Werner, 1986: Simultaneous Submillimeter and Infrared Observations of Flat-Spectrum Radio Sources, Astrophys. J., 304, 646.

Rosenkranz, P.W., 1975: Shape of the 5 mm Oxygen Band in the Atmosphere, IEEE Trans. Antennas Propag., AP-23(4), 498.

Rothman, L.S., R.R. Gamache, A. Goldman, L.R. Brown, R.A. Toth, H.M. Pickett, R.L. Poynter, J.-M. Flaud, C. Camy-Peyret, A. Barbe, N. Husson, C.P. Rinsland, and M.A.H. Smith, 1987: The HITRAN Database: 1986 Edition, Appl. Opt., 26, 4058.

Smith, L.L., and T. Hilgeman, 1981: High-Resolution Lower Atmospheric Transmission Predictions Over Long Paths, in Atmospheric Transmission (R.W. Fenn, ed.), Proc. SPIE 277, 141, Intl. Soc. for Optical Engineering, P.O. Box 10, Bellingham, WA 98227.

Whitcomb, S.E., R.H. Hildebrand, J. Keene, R.F. Stiening, and D.A Harper, 1979: Submillimeter Brightness Temperatures of Venus, Jupiter, Uranus, and Neptune, Icarus, 38, 75.

RADIATIVE TRANSFER AND RETRIEVAL THEORY FOR THE 183-GHZ WATER AND OZONE CHANNELS ON MLS

W.A. Lahoz and R.S. Harwood
Department of Meteorology, Edinburgh University
Edinburgh EH9 3JZ, UK

ABSTRACT

The Microwave Limb Sounder (MLS) to be carried on the Upper Atmosphere Research Satellite (UARS) will measure concentrations of a number of stratospheric constituents. In this paper, sample calculations of the forward problem pertaining to the 183-GHz band of MLS are presented. This band is used to infer water vapour and ozone amounts from the 183.3-GHz and 184.4-GHz lines, respectively. The calculations include the effects of the antenna field of view and radiometric filter response.

Some indications of the sensitivity of the calculated radiances to changes in the constituent concentration are shown. These are closely related to the "weighting functions" which arise in the "retrieval" or "inverse problem."

1. INTRODUCTION

The MLS measures atmospheric thermal emission from selected molecular spectral lines at millimetre wavelengths (NASA, 1985). From the intensity and spectral characteristics of this emission and its variation as the MLS field of view (FOV) is scanned vertically through the atmospheric limb, profiles of geophysical parameters are inferred. The MLS is being developed by a consortium of groups at the Jet Propulsion Laboratory (JPL) in the USA and at the Rutherford-Appleton Laboratories and Heriot-Watt and Edinburgh Universities in the UK, under the overall direction of J. Waters at JPL. The principal UK contribution is to the 183.3-GHz and 184.4-GHz channels. These will allow the concentration of these species to be determined in the mesosphere to greater heights than would otherwise be possible. Several scientific investigations using MLS data are envisaged: determination of the daily, global distributions and climatology of ozone and water vapour, including seasonal variations (and interannual ones if the instrument's lifetime permits); investigation of the ozone chemistry in the upper stratosphere and mesosphere; radiation budget studies of the middle atmosphere; dynamical studies of the circulation of the middle atmosphere; validation programs with other measurements.

MICROWAVE REMOTE SENSING
of the EARTH SYSTEM
Alain Chedin (Ed.)

To understand the radiation that will be measured by the satellite, a thorough study of the forward problem--viz. the solution of the equation of radiative transfer--is needed. This we do by choosing several algorithms and keeping each algorithm as general as possible. In this manner we are able to perform inter-algorithm comparisons. This philosophy has been extended to the derivation of antenna and filter brightness temperatures, which take account of the effect of the antenna (FOV) and filter responses.

The retrieval scheme to be implemented--viz. sequential estimation (Froidevaux et al.;Rodgers, 1976)--involves the so-called weighting functions. These can be calculated by means of the forward problem. Clearly, a forward problem which is as accurate as possible is essential for this exercise.

In this paper, section 2 concerns our treatment of the forward problem; section 3 deals briefly with the evaluation of the weighting functions necessary for the retrieval scheme, and section 4 deals with future work.

2. FORWARD PROBLEM

The radiative transfer equation provides the mathematical description of atmospheric emission and absorption (Chandrasekhar, 1960). For nonscattering, nonrefractive absorption at microwave frequencies, the scalar radiative transfer equation has the form (Waters,1976) (cf. the tensor radiative transfer equation in Lenoir, 1967; for H_2O and O_3 the scalar equation applies):

$$I_\nu(s)=I_\nu(0)exp(-\tau_\nu(0,s))+\int_0^s B_\nu(T)exp(-\tau_\nu(s',s))k_\nu(s')ds' \tag{1}$$

where $I_\nu(s)$ is intensity of radiation at frequency ν and position s, $B_\nu(T)$ is the Planck function giving the intensity of radiation at frequency ν from a black body at temperature T, $k_\nu(s)$ is the absorption coefficient per unit length at frequency ν and position s and

$$\tau_\nu(s',s)=\int_{s^1}^s k_\nu(s'')ds'' \tag{2}$$

is the optical path between points s and s'.

The Planck function is given by:

$$B_\nu(T)=2h\nu^3[exp(h\nu/kT)-1]^{-1}/c^2 \tag{3}$$

For microwave frequencies and atmospheric temperatures $h\nu/kT<<1$ and we can approximate $B_\nu(T)$ with sufficient accuracy by expanding the exponential and ignoring terms of $O \sim (h\nu/kT)^4$.

Defining a brightness temperature T_B (units of temperature) by $T_B \equiv c^2 I_\nu / 2k\nu^2$ we can rewrite Eq. (1) as:

$$T_B(s) = T_B(0)\exp(-\tau_s) + \int_0^{\tau_s} B_\nu^*(T)\exp(-\tau_\nu(s',s))d\tau_\nu(s',s) \qquad (4)$$

where $T_B(0) = 2.7$ K is the background radiation, we have defined $B_\nu^*(T) \equiv c^2 B_\nu(T)/2k\nu^2$, $\tau_s \equiv \tau_\nu(0,s)$ and made an appropriate change of variables. The evaluation of T_B from Eq. (4) given a particular temperature profile and distribution of absorbers is called the forward problem. In practice one knows the radiation profile and wishes to obtain the relevant geophysical parameters; this is the retrieval problem.

Solution of Eq. (4) is performed by dividing the atmosphere into spherical layers, not necessarily of the same thickness, and assuming that the atmosphere is spherically symmetric. As this assumption is built into the retrieval scheme to be implemented we incorporate it into the algorithms which evaluate Eq. (4). With this assumption, calculation of the brightness temperature contribution from the atmosphere is considerably simplified.

The integrals concerned with the equation of radiative transfer can be approximated by splitting the atmosphere into layers (l=1,L) giving:

$$T_B(s) = T_B(0)\exp(-\tau_s) + \sum_{\ell=1}^{L} [B_\nu^*(T)\exp(-\tau_\nu(s',s))\Delta\tau_\nu(s',s)]_l \qquad (5a)$$

$$[\tau_\nu(s',s)]_l = \sum_{i \leq \ell' \leq \ell} [k_\nu(s'')\Delta s'']_l \qquad (5b)$$

with $\Delta s''$ obtained from the geometry of the system. Alternatively, one can split the integral in Eq. (4) over the atmospheric layers. By assuming a particular parametric dependence for $B_\nu^*(T)$ over a layer, one may evaluate the integral analytically. The relevant algorithms have been coded as they apply to the 183-GHz band. In the calculations discussed in this paper, we assume a satellite height of 600 km giving a distance of approximately 2700 km from UARS to the tangent point at 35 km in height.

The algorithms developed take account of the fact that either pressure or geometric height can be used as the height variable. With this choice the following algorithms have been developed: (i)divide the atmosphere into 20×2^n layers of constant mixing ratio and temperature, each of thickness $5/2^n$ km (n=0,1,...); (ii)consider divided and undivided atmospheric paths from the top of the atmosphere to the tangent point and compare calculations with respect to a predetermined criterion: if criterion is not met, the path is subdivided and procedure repeated; (iii)as (ii) but treating one by one the layers along the radiation path, rather than the whole path; (iv)an interactive scheme where divided and undivided paths are considered but the integration limits of the radiation path are allowed to vary[1]. Further algorithms based on analytical results available under certain conditions are being developed with the aim of providing checks to the numerical algorithms described above.

The effect of the instrument finite FOV is taken into account by introducing a so-called antenna brightness temperature $T_{B,\nu}{}^A$, defined to be the normalised convolution of the antenna apparatus function and the true (monochromatic) atmospheric brightness temperature, which is given by Eq. (4):

$$T_{B,\nu}{}^A \equiv \frac{\int A(\theta_\alpha - \theta) T_{B,\nu}(\theta) d\theta}{\int A(\theta_\alpha - \theta) d\theta} \tag{6}$$

$$A(\theta_\alpha - \theta) = 0 \text{ for } \theta > \theta_\alpha + \theta_n, \ \theta < \theta_\alpha - \theta_n$$

In Eq. (6) θ_α is the antenna angle and θ is the ray angle, and the $T_{B,\nu}$ vertical distribution has been changed to an angular one. $A(\theta)$ is the antenna apparatus function which takes account of the finite FOV and smears out the $T_{B,\nu}$ angular distribution (Brault and White, 1971). In writing down Eq. (6) we have assumed $A(\theta)$ does not have horizontal variation and that angular displacements from the origin of the calculation (antenna mid-angle) are small. We use Eq. (6) in our antenna brightness calculations with $A(\theta)$ a truncated Gaussian set to zero where the fall off is greater than 15 db.

Similarly, filter effects are taken into account by introducing a so-called filter brightness temperature $T_{B,\nu}{}^F$ defined to be the normalised convolution of the filter response $F_i(\nu)$ and $T_{B,\nu}{}^A$. For filter (channel) i, which ideally measures the radiation falling in frequency interval $[\nu_i, \nu_{i+1}]$, we have:

$$[T_{B,\nu}{}^F(\theta)]_i \equiv \frac{\int T_{B,\nu}{}^A(\theta) F_i(\nu) d\nu}{\int F_i(\nu) d\nu} \tag{7}$$

where the contributions from the image sidebands are ignored due to being negligible with respect to the sidebands of interest (Froidevaux et al. ibid.).

[1]   Read, W. G., 18 January 1988: Upper Atmosphere Research Satellite (UARS) Microwave Limb Sounder (MLS) Direct Measurement Model for Radiative Spectral Intensity Version 2.1, JPL Document.

Our algorithms have been designed to handle four kinds of filter response: box-car, linear, quadratic, and shapes expected from JPL data[2]. The MLS will have 15 filters (channels) each for detecting radiation from the 183.3-GHz H_2O and 184.4-GHz O_3 lines. Filter 8 will be centred at line centre whereas filters 1 and 15 will be centred at -191 MHz and +191 MHz from line centre respectively.

The inputs in the monochromatic calculations are as follows. The spectroscopic parameters are from Waters (ibid.) for H_2O and from Gora (1959) for O_3; we also use data from the JPL spectral line catalogue (Poynter and Pickett, 1984). For line shape we use the Voigt profile (Drayson, 1976), although the algorithms can employ a Lorentz (Lorentz, 1906), Van Vleck-Weisskopf (VVW) (Van Vleck and Weisskopf, 1945), or Doppler (Waters ibid.) line shape. Calculations done by the authors suggest that the difference between convolving a Doppler with a Lorentz profile and convolving a Doppler with a VVW to obtain the Voigt profile is completely negligible. The temperature profiles are obtained from Houghton (1986) and the mixing ratio profiles (Figs. 1,4) for our "standard" runs, which are typical of mid-latitude values. We also perform calculations with a wet upper troposphere (constant water vapour mixing ratio of 1000 ppmv), because 1 part in 1000 is very dry for surface level air. To ascertain the impact of different mixing ratio profiles in the upper stratosphere and mesosphere, we carry out calculations where we change these "standard" profiles. For both H_2O and O_3 we include the effects of other water lines between $\approx$ 22 GHz and $\approx$ 1000 GHz (water lines with frequency >1000 GHz are sufficiently far away to be ignored) and the water continuum by means of the Gaut-Reifenstein correction (Gaut and Reifenstein, 1971). Calculations done by the authors (Figs. 1-6) indicate that these effects should be included only for tangent heights below $\approx$ 30 km. This cut-off point is chosen at a height well above the region of the atmosphere, where their effects are significant to minimise the error involved in ignoring these effects. Figure 1 refers to a calculation of monochromatic brightness temperature at centre of channels 1-8 against the tangent pressure of the centre of the FOV for H_2O in which no extra effects have been included. Figure 2 is as Fig. 1 but with liquid water, extra water lines, and the Gaut-Reifenstein correction added. Figure 3 is as Fig. 1 but with a very wet upper troposphere (otherwise the H_2O mixing ratio profile is as for Fig. 1). Figure 4 is as Fig. 1 except that the calculation is for ozone. Figure 5 is as Fig. 4 but with other water lines (assuming a wet upper troposphere), and the Gaut-Reifenstein correction included. Figure 6 is as Fig. 5 except that we include liquid water.

[2] JPL data from filter bank tests, 1986.

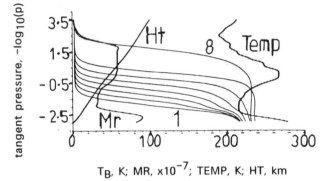

TB, K; MR, x10^{-7}; TEMP, K; HT, km

Figure 1. Plot of brightness temperature (T$_B$) versus tangent pressure
(-log$_{10}$(p); p in mbars). The calculation is done at the centre
of channels 1-8 of the 183.3-GHz H$_2$O line, with no extra effects
included and using scheme (i). We have superimposed plots of
temperature (TEMP) and mixing ratio (MR) profiles used in the
calculation, as well as a plot of tangent height (HT), versus
tangent pressure.

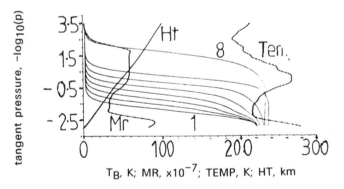

TB, K; MR, x10^{-7}; TEMP, K; HT, km

Figure 2. As fig. 1 but with liquid water, extra water lines and the water
continuum included.

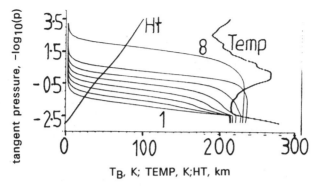

Figure 3. As fig. 1 but with a constant water vapour mixing ratio of 1000 ppmv in the troposphere.

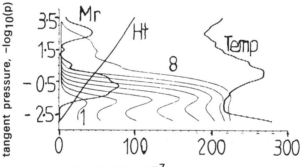

Figure 4. As fig. 1, except that the calculation is for the 184.4-GHz O_3 line.

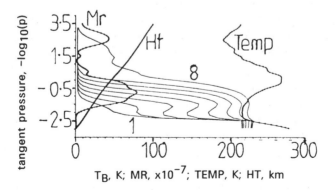

Figure 5. As fig. 4 but with other water lines (assuming 1000 ppmv of water vapour in the troposphere) and water continuum included.

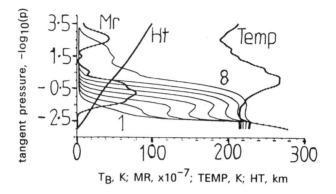

Figure 6. As fig. 5, except that we include water.

To study the effect of liquid water on the 183-GHz band measurements, we parametrise the concentration of liquid water with respect to temperature using data available (e.g., Fei'gelson, 1966) which is either typical of average values or is averaged to take account of different types of cloud. We employ Eq. (22b) in Liebe (1981) to represent the liquid water contribution to the absorption coefficient. Calculations that include liquid water and other water lines show that their effect is very small (~0.5%) for 183.3-GHz H_2O and significant (~5-10%) for 184.4-GHz O_3 (Figs. 4,6). The effect of liquid water on O_3 depends on how wet the troposphere is. For a constant mixing ratio for water vapour of 1000 ppmv the effect is negligible; however, for a dry troposphere (see Fig. 1), the effect is significant (~5-10%), but only for tangent heights below $\approx$ 10 km, where meaningful measurements probably will not be possible. Consequently, although liquid water must be taken into account, it is not necessary to employ highly accurate parametrisations to do so. We note that these calculations suggest that measurements of liquid water may be possible using the 184.4-GHz O_3 line, although this will depend on the amount of water vapour in the troposphere.

When calculating absorption coefficients, spectroscopic parameters such as line width, temperature exponent and line strength appear. It is important that these are known accurately. In order to ascertain their impact, sensitivity tests have been carried out (see Figs. 7-12), in which the relevant spectroscopic parameters have been changed by 10% and the monochromatic brightness temperature recalculated. In these calculations the effects of the water continuum and liquid water have been neglected. The mixing ratio profiles are the "standard" ones (cf. Figs. 1,4), the temperature profile is appropriate to mid-latitude, and the algorithm used is scheme (iii). The results suggest that the most critical spectroscopic parameter is line intensity (see Figs. 9,12) and that errors in it will give comparable errors in the mixing ratio. At the moment line intensities are known to ~5%, but it appears possible to improve their accuracy (in particular to ~1% for H_2O; private communication from H. Pickett, December 1987). We note the presence of spikes in these plots; they are due to the

temperature and mixing ratio profiles not being monotonic. For a proper evaluation of the quantitative effects of errors in the spectroscopic parameters, the finite FOV and filter convolutions must be included. This work is currently under way.

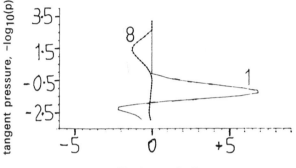
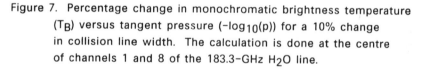

Figure 7. Percentage change in monochromatic brightness temperature
(T_B) versus tangent pressure ($-\log_{10}(p)$) for a 10% change
in collision line width. The calculation is done at the centre
of channels 1 and 8 of the 183.3–GHz H_2O line.

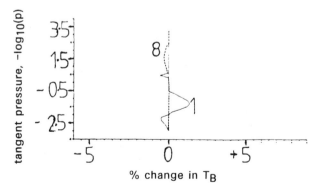
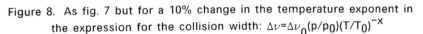

Figure 8. As fig. 7 but for a 10% change in the temperature exponent in
the expression for the collision width: $\Delta\nu = \Delta\nu_0 (p/p_0)(T/T_0)^{-x}$

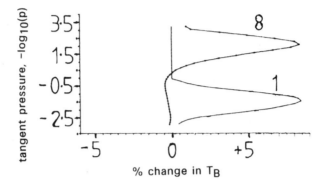

Figure 9. As fig. 7 but for a 10% change in the line strength.

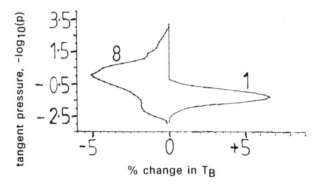

Figure 10. As fig. 7 but for the 184.4–GHz O_3 line.

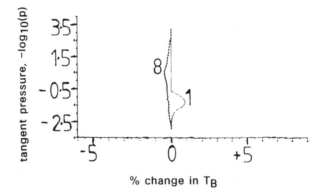

Figure 11. As fig. 8 but for the 184.4–GHz O_3 line.

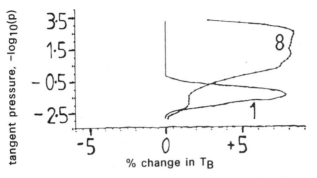

Figure 12. As fig. 9 but for the 184.4-GHz O_3 line.

It is well known that the Voigt line shape contains an inconsistency--viz., that one assumes the Lorentz half-width to be independent of the Doppler velocity (Drayson, ibid.)--when the Lorentz half-width depends on temperature and thus has a correlation with the Doppler velocity. Despite this, use of the Voigt line shape can be justified on the grounds that the Lorentz half-width depends weakly on temperature and that including this velocity dependence would lead to unacceptably complicated computations. Although more complicated line shapes have been derived (Galatry, 1961), it appears that the most fruitful improvement will come from velocity averaging (Pickett, 1980).

One of the advantages of microwave remote sensing with respect to infra-red remote sensing is that the assumption of local thermal equilibrium (LTE) holds at higher altitudes. From the point of view of radiative transfer, non-LTE in the MLS 183-GHz band can safely be discounted on the grounds that both the H_2O and O_3 lines are rotational transitions in the vibrational ground state and that the nearest lines in excited vibrational states are more than 200 MHz away and extremely weak (Kerridge, personal communication January, 1988). Furthermore, assuming LTE, the contribution to the absorption coefficient of all lines from the H_2O ν-2 state with frequency less than 1000 GHz is negliglible (<0.01%) in comparison with the contribution of all H_2O lines in the ground state with frequency < 1000 GHz for tangent heights less than 100 km. Similarly, the contribution to the absorption coefficient of all lines within 5 GHz of the 184.4-GHz O_3 line from the O_3 excited ν_1, ν_1, ν_3 and $2\nu_2$ is negligible for tangent heights $10 < h < 100$ km. For tangent heights this contribution is ~1%; however, at these tangent heights the effects of water continuum, liquid water and tropospheric water vapour will swamp the contributions from the excited states. From the point of view of the retrieval , one is still retrieving the number of molecules in the ground vibrational state.

The mixing ratios of O_3 and H_2O are not very well known in the upper stratosphere and mesosphere. In view of this, it is important to investigate how changes in the mixing ratio of these species affect the radiation incident on the MLS. For reasons of economy we investigate six scenarios, three per species. For H_2O we have a "standard" scenario (Fig. 1), but with a constant mixing ratio of 1000 ppmv in the troposphere, and scenarios where there is more H_2O (Fig. 13) and less H_2O (Fig. 14) in the middle to upper atmosphere than in the "standard" scenario. The results suggest that, as expected, the off-centre channels, are relatively unaffected. The centre channel , however, is significantly affected. From these calculations, depending on the mixing ratios in the upper atmosphere, one can measure H_2O from ~80-~100 km to ~15 km for an integration time of ~ 1 second, and higher for longer integration times obtained by averaging observations.

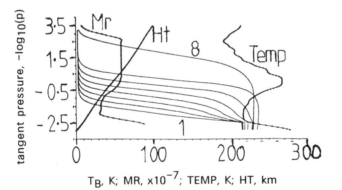

Figure 13. As fig. 3 but with more water vapour in the upper atmosphere.

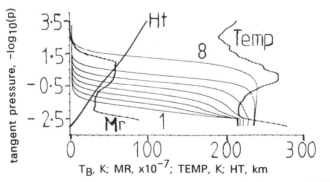

Figure 14. As fig. 3 but with less water vapour in the upper atmosphere.

For O_3 we have a "standard" night scenario (Fig. 4) and scenarios where there is a sharp maximum in the middle-stratosphere and little O_3 in the middle to upper atmosphere (Fig. 15), and where the distribution of O_3 in the middle stratosphere follows a plateau, with little O_3 above (Fig. 16). In these calculations we include the effect of other water lines and the water continuum but neglect liquid water. The results are qualitatively similar to those for H_2O. From these calculations, depending on the mixing ratios in the upper stratosphere, one can measure O_3 from ~75-~95 km to ~20 km for an integration time of ~ 1 second, and higher for longer integration times obtained by averaging observations.

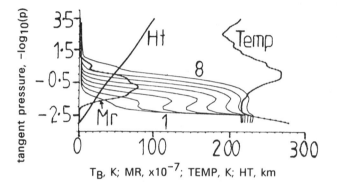

Figure 15. As fig. 5 but with less O_3 in the upper atmosphere.

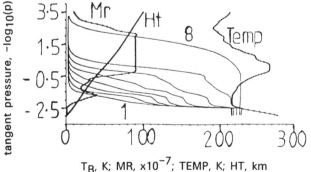

Figure 16. As fig. 5 but with the O_3 distribution following a plateau in the middle atmosphere and falling off in the upper atmosphere.

For a proper evaluation of how changes of mixing ratio in this atmospheric region affect the forward problem, one must include the finite FOV and filter convolutions. Consequently, the above height ranges and plots must be compared with those for filter brightness temperature (Figs. 17-21).

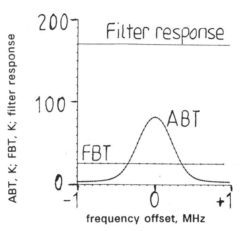

Figure 17. Plot of frequency offset (MHz) versus antenna brightness temperature (ABT) and filter brightness temperature (FBT). The calculation is for channel 8 of the 183.3-GHz H_2O line and at a tangent height of 80 km. It uses a box-car function filter response and scheme (i).

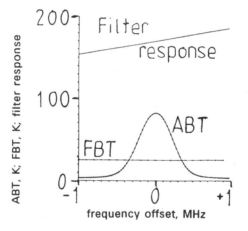

Figure 18. As fig. 17 but with a linear filter response.

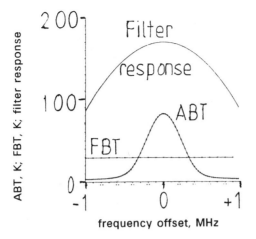

Figure 19. As fig. 17 but with a quadratic filter response.

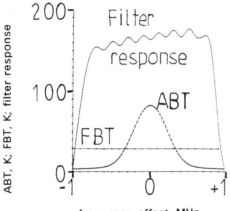

Figure 20. As fig. 17 but with the filter response shown in the plot.

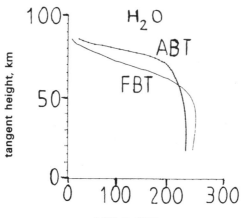

ABT, K; FBT, K

Figure 21. Plot of antenna brightness temperature (ABT) and filter
brightness temperature (FBT) versus tangent height. ABT is
calculated at the centre of channel 8, and FBT is calculated
for channel 8. The calculation uses the filter response in
fig. 20 and scheme (iii).

The above calculations have been performed assuming no Doppler shift.
The maximum combined velocity of earth, UARS, and wind is considerably
less than 1000 m/s, which gives a maximum Doppler shift considerably less
than 1 MHz. This suggests that the filter brightness temperature associated
with the far-wing channels is likely to be unaffected. The effect on other
channel filter brightness temperatures will depend on the slope of the curve
of filter brightness temperature versus frequency (frequency as abscissa) and
channels (filter) width. Given that the Doppler velocity should be
determinable from the difference between channels i and 16−i (i=1,...,7), this
result suggests that only middle and off-centre channels will be suitable for
determining it. The effect of a Doppler shift of 1000 m/s on filter 8 for O_3
can be ascertained by comparing Figs. 22−23. The effect for O_3 is 0∼10% at
most.

Effects of refraction and the earth's reflectivity[3] have not been included in
these calculations. They will be included in future work.

3   Read, ibid.

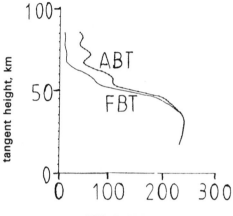

Figure 22. As fig. 21 but for the 184.4–GHz O_3 line.

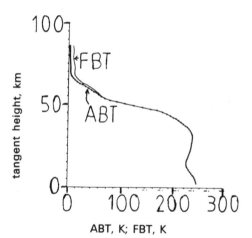

Figure 23. As fig.22 but including a Doppler shift of 1000 m/s.

3. RETRIEVAL SCHEME

The retrieval scheme to be implemented in the MLS UARS experiment is sequential estimation (Rodgers, 1976). Central to this scheme are the so-called weighting functions $\underset{\sim}{K}$ (Rodgers ibid.; Froidevaux et al. ibid.). From their definition these can be calculated by:

$$K_{ij} = \frac{\partial (T_{B,\nu}{}^F)_i}{\partial x_j} \Big|_{\vec{x}=\vec{x}} \qquad (8)$$

where $\vec{x}$ refers to temperature, mixing ratio or any other parameter, $T_{B,\nu}{}^F$ is filter brightness temperature (cf. Eq. (7)), and the subscripts refer to the relevant region of the atmosphere.

In practice, K_{ij} will be calculated numerically by:

$$K_{ij} \approx [(T_{B,\nu}{}^F(x_1,...,x_j+\delta x_j,...))_i - (T_{B,\nu}{}^F(x_1,...,x_j,...))_i]/\delta x_j \qquad (9a)$$

or,

$$K_{ij} \approx [(T_{B,\nu}{}^F(x_1,...,x_j+\delta x_j,...))_i - (T_{B,\nu}{}^F(x_1,...,x_j-\delta x_j,...))_i]/2\delta x_j \qquad (9b)$$

at evenly spread log pressure surfaces. Consequently, the subscript in Eq. (9), referring to the vector components of $\vec{x}$, will refer to a particular pressure surface.

In view of the asumption of linearity built into the retrieval scheme, sufficient $\underset{\sim}{K}$'s must be chosen such that the retrieval problem remains locally linear. A full-scale calculation to determine how many $\underset{\sim}{K}$'s will be needed is not available yet; however, filter brightness temperature calculations in which the mixing ratio profile remains constant but the temperature profile is changed drastically (e.g. polar instead of equatorial temperature profile) have been carried out and the difference calculated (Fig. 24). Essentially, this gives an idea of the error in employing the wrong latitudinal temperature profile. To get a percentage error for,say, filter 8, these plots must be compared with those in (Figs. 17-21). These suggest that, for filter 8, the error for an O_3 night profile (see Fig. 4) is $\leq$ 5% below tangent heights $\approx$ 75 km. Some sample weighting functions, calculated for a 10% change in mixing ratio at four different tangent heights are shown in Fig. 25. The smearing due to the FOV is clearly evident.

4. FUTURE WORK

Future pre-launch work will concern algorithm validation employing UARS ephemeris data and other data/effects not included in the calculations described in this paper, but which have to be included when calculating the MLS weighting functions. Subsequently, the above-mentioned weighting functions will be computed, and calculations will be carried out to ascertain the number of weighting functions that will be required to carry out the retrieval scheme satisfactorily.

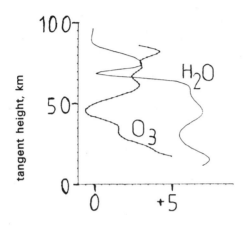

delta FBT, K

Figure 24. Plot of difference in filter brightness temperature (DFBT) versus tangent height. The calculation is for the difference in FBT between assuming a polar temperature profile and an equatorial temperature profile, the mixing ratio profile remaining unchanged. It is performed for the 183.3-GHz H_2O line and the 184.4-GHz O_3 line, and uses scheme (i) for H_2O and scheme (iii) for O_3.

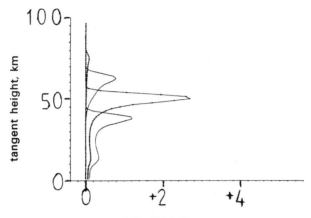

Figure 25. Plot of the change in filter brightness temperature (DFBT) versus tangent height for a 10% change in mixing ratio in one layer at four different heights (evident from where the curve maxima lie). The calculation is for channel 8 of thej 184.4–GHz O_3 line. The mixing rtaio profile is as fig. 4.

ACKNOWLEDGMENT

This work was supported by a research grant from the Science and Engineering Research Council of Great Britain.

REFERENCES

Brault, J. W., and O. R. White, 1971: The Analysis and Restoration of Astronomical Data via the Fast Fourier Transform, Astron. & Astrophys., 13, 169-189

Chandrasekhar, C., 1960: Radiative Transfer, Dover Publications, New York, 393pp.

Drayson, S.R., 1976: Note Rapid Computation of the Voigt Profile, J. Quant. Spectrosc. Radiat. Transfer, 16, 611-614.

Fei'gelson, E. M., 1966: Light and Heat Radiation in Stratus Clouds, Program for Scientific Translations, Ltd. (Translated from Russian), 245pp.

Froidevaux L., W. G. Read, J. W. Waters, and W. A. Lahoz (Ed. W. A. Drew), 7 July 1987: Upper Atmosphere Research Satellite (UARS) Microwave Limb Sounder (MLS) Level-2 Data Processing: Theoretical Basis, JPL Document L2DP TB 1.4.

Galatry, L., 1961: Simultaneous Effect of Doppler and Foreign Gas Broadening on Spectral Lines, Phys. Rev., 122, 1218-1223.

Gaut, N. E., and E. C., Reifenstein III, 1971: Environmental Res. and Tech. Rep. No. 13, Lexington, Massachusetts.

Gora, E. K., 1959: J. Mol. Spectrosc., 3, 78.

Houghton, J. T., 1986: The Physics of Atmospheres, 2nd Edition, Cambridge University Press, Cambridge, 271pp.

Lenoir, W. B., 1967: J. App. Phys., 38, 5283-5290.

Liebe, H. J., 1981: Modeling Attenuation and Phase of Radio Waves in Air at Frequencies Below 1000 GHz, Radio Science, 16, 1183-1199.

Lorentz, H. A., 1906: Proc. Ams. Sci., 8, 59.

NASA Document 430-1003-001, 1985: Upper Atmosphere Research Satellite (UARS) Mission (Available from Goddard Space Flight Center, Greenbelt, Maryland).

Pickett, H. M., 1980: Effects of Velocity Averaging on the Shapes of Absorption, J. Chem. Phys., 73, 6090–6094.

Pickett, H. M., 1987: Private Communication.

Poynter, R.L., and H. M. Pickett, 1984: Submillimeter, Millimeter and Microwave Spectral Line Catalog, JPL Publication 80–23, Revision 2.

Rodgers, C. D., 1976: Retrieval of Atmospheric Temperature and Composition from Remote Measurements of Thermal Radiation, Rev. Geophys. & Sp. Phys., 14, 609–624.

Van Vleck, J.H., and V. F. Weisskopf, 1945: On the Shape of Collision–Broadened Lines, Rev, Mod. Phys., 17, 227–236.

Waters, J. W., 1976: Absorption and Emission by Atmospheric Gases. In M.

L. Meeks (Ed.), Methods of Experimental Physics, 12B, Academic Press, New York, 142–176.

SIMULATION OF THE EFFECT OF A REALISTIC CLOUD FIELD ON THE AMSU MEASUREMENTS

H. Legleau
Centre de Météorologie Spatiale
22302 Lannion BP147 France

ABSTRACT

The measurements in the different channels of the future AMSUA (advanced microwave sounding unit A) and AMSUB sensors have been simulated over AVHRR sceneries, with a microwave radiative transfer model (using a cloud field derived from the AVHRR (advanced very high resolution radiometer) channels 2 and 4, and a given atmospheric profile), combined with a geometrical simulation of the field of view (FOV) of the instruments (i.e, the location of the centers of the FOV and the spatial response of the antenna). This simulation has been used to study the effect of a realistic cloud field on the microwave AMSUA and AMSUB measurements:

It was found that clouds may have important effects on AMSU brightness temperatures, especially water clouds on AMSUA and AMSUB over sea, and ice clouds on AMSUB over land and sea.

It has been shown that, over sea, the contamination due to clouds in the window channels 2(31 Ghz), 3(50 Ghz) and 15(89 Ghz) of the AMSUA sensor can be regressed from the contamination in the window channel 1(23 Ghz) of the same sensor, irrespective of the scanning angle, the cloud type and the vertical structure of the atmosphere. This can be explained by the low value of the emissivity of sea and the rather weak importance of the scattering process in clouds at those frequencies.

It is shown that it should be possible to use those "cloud contamination" regressions to check the presence of precipitation, and to get the clear brightness temperatures of the microwave window channels of the AMSUA sensor in areas without rain.

1.   INTRODUCTION

The next generation of NOAA polar orbiting satellites will have a new microwave sounding unit (the AMSU instrument), which will be used for temperature sounding (AMSUA) and water vapor content sounding (AMSUB) in cloudy regions opaque to infra-red radiations.

MICROWAVE REMOTE SENSING
of the EARTH SYSTEM
Alain Chedin (Ed.)

Microwave channels are usually supposed not to be
affected by clouds (the present MSU processing assumes no
contamination due to clouds).

This work describes a simulation of the effects of a
realistic cloud field (derived from AVHRR sceneries) on
the AMSU brightness temperatures.

2.   DESCRIPTION OF THE AMSUA-AMSUB INSTRUMENTS

2.1   TECHNICAL AND PHYSICAL CHARACTERISTICS

The technical characteristics of the AMSU have been
drawn from Pick (1985) and Calhon (1984). The shape and
width of the field of view (FOV) are described in Fig. 2
and table 2. (The internal ellipse is the half-beam footprint, whereas the external ellipse is the FOV footprint
i.e., the surface contributing to 98% of the measured value). We can notice the rather large FOV footprints of
channels 166 Ghz and 183 Ghz of the AMSUB, due to the
side lobes.

The main physical characteristics of the 15 AMSUA
channels and the 5 AMSUB channels (i.e., the major absorbing gas, the pressure of the peak of the weighting function and the transmittance for the U.S standard atmosphere) are summarized in table 1. The polarization of the
different channels was not decided when that work was done; they are all supposed unpolarized.

2.2   THE AMSU AND THE CLEAR ATMOSPHERE

This has already been summarized in table 1, with
the nature of the major absorbing gas, the value of the
total transmittance and the height of the peak of the
weighting function for each AMSU channel.

2.3   THE AMSU AND THE SURFACE EMISSIVITY

The emissivity of land, which depends heavily on the
type and the moisture of the ground, will be taken equal
to 0.95 in this simulation. The emissivity of sea depends
mainly on frequency, polarization and wind speed, as
shown on Fig. 1 a) and Fig. 1 b).

2.4   THE AMSU AND CLOUDS AND RAIN

The water clouds are only an absorbing and emitting
medium for the AMSUA and AMSUB frequencies (no scattering)); hence, the effect of water clouds on AMSU does
not depend on the particle size distribution.

TABLE 1. CHARACTERISTICS OF THE AMSU CHANNELS.

channel number	frequency (GHZ)	pressure at peak of temperature weighting fonction (U.S atm. stand.)	major absorbing gas and totale transmittance (U.S atm. stand.)		noise (degree)
AMSUA					
1	23.8	surface	H2O	.9	0.2/0.5
2	31.4	surface	H2O	.93	0.2/0.5
3	50.3	surface	O2 AND H2O	.65	0.3/0.4
4	52.8	surface	O2	.3	0.25
5	53.596 +/-.115	700 mb	O2	.12	0.25
6	54.4	400 mb	O2	.02	0.25
7	54.94	270 mb	O2	.002	0.25
8	55.5	180 mb	O2	.00C1	0.25
9	57.290344	90 mb	O2	.0	0.25
10	+/- .217	50 mb	O2	.0	0.35
11	+/- .3222 +/- .048	25 mb	O2	.0	0.35
12	+/- .022	12 mb	O2	.0	0.5
13	+/- .01	5 mb	O2	.0	0.8
14	+/- .0045	2 mb	O2	.0	1.3
15	89.0	surface	H2O	.82	0.6/1.0
AMSUB					
16	89.0	surface	H2O	.82	0.6/1.0
17	166.0	surface	H2O	.48	0.6/1.0
18	183.31 +/- 1.0	440 mb	H2O	.0	0.6/1.0
19	+/- 2.0	600 mb	H2O	.001	0.6/1.0
20	+/- 3.0	800 mb	H2O	.09	0.6/1.0

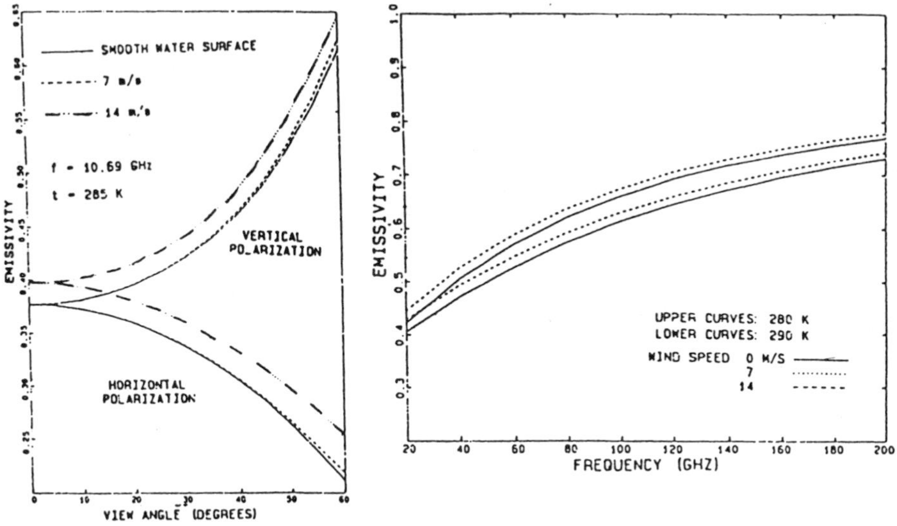

Figure 1. Sea emissivity (after Warner,C.,1986 personal communication).

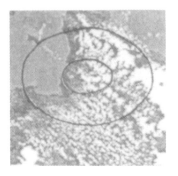
AMSUA footprint   AMSUB 89 Ghz footprint

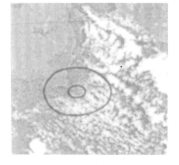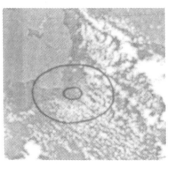

AMSUB 166 Ghz footprint   AMSUB 183 Ghz footprint

Figure 2. Shape of the different AMSUA and AMSUB Field Of
         View footprints used in our simulation on an
         AVHRR image . The internal ellipse is the "half
         beam" footprint; the external ellipse is the
         FOV footprint.

TABLE 2. SIZE OF THE FIELD OF VIEW FOOTPRINTS OF THE
        AMSUA AND AMSUB SENSORS USED IN OUR SIMULATION.

frequency instrument	number of FOV per line	angle half-beam/FOV (degree)
AMSUA	30	3.3 / 8.6
AMSUB 89 Ghz	90	1.1 / 2.4
AMSUB 166 Ghz	90	1.1 / 4.4
AMSUB 183 Ghz	90	1.1 / 5.2

The ice clouds have practically no effects on AMSUA channels (except at 89 Ghz) and are essentially a scattering medium for AMSUB frequencies; hence, the effect of ice clouds on AMSUB depends strongly on the size of the particles.

The rain is an absorbing, emitting, and scattering medium for AMSUA and AMSUB frequencies; hence, the effect of rain on AMSU depends strongly on the droplets size distribution.

3. DESCRIPTION OF THE SIMULATION

This simulation, which is done only during the day, is the study of the effects of clouds on the brightness temperature field of the future microwave sensors AMSUA and AMSUB over a whole orbit.

The usual simulations consist only in a radiative transfer through a given cloud. This simulation emphasizes the horizontal aspect in two ways: a realistic cloud field, derived from AVHRR data, is used in an horizontally homogeneous atmosphere, and the spatial response of the instrument (the antenna pattern) is taken into account. The simulation is done in five steps.

3.1 THE DIFFERENT STEPS OF THE SIMULATION

The cloud parameters (the top temperature, the total liquid water content (LWC) and the fractional cloudiness) are first deduced from the AVHRR channels 2 (visible) and 4 (infra-red), using a scheme described by Saunders (1986, 1988) and Kriebel (1986). This is done for every AVHRR pixels.

Then, the vertical structure of the clouds is obtained for each AVHRR pixel; some assumptions have been made, derived from Warner (1984) and from Pruppacher and Klett (1978):
-the atmosphere is horizontally homogeneous over the whole orbit (i.e., a single radiosonde is used).
-the ratio ice/water within the cloud varies linearly with the temperature, from 0% at 268°K to 100% at 243°K.
-the density of the water clouds is 0.25g/m3.
-the density of the ice clouds is 0.10g/m3.
-the diameter of the largest ice particles is 3 mm; the concentration of the ice particles decreases logarithmically with the size.

The computation of the microwave brightness temperature for all the AMSU frequencies is done for each AVHRR pixel through the clouds computed at the previous step, using the radiative transfer model developped by C.Warner

Figure 3. Atlas map corresponding to the maps of contamination by clouds presented in Fig. 7-8.

Figure 4. Map of AVHRR channel 4 corresponding to the maps of contamination by clouds presented in Fig. 7-8. The AMSUA half-beam footprints are are superimposed.

Figure 5. Example of a map of the contamination by clouds
          of the AMSUA/89 Ghz (DTB(15)):
          DTB(15) = TB(15) − TBcl(15)   in 1/100 °K

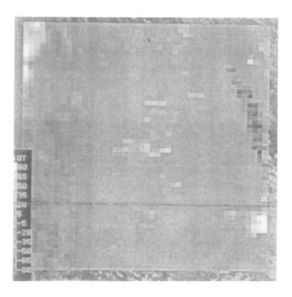

Figure 6. Example of a map of the contamination by clouds
          of the AMSUB/89 Ghz (DTB(16)):
          DTB(16) = TB(16) − TBcl(16)   in 1/10 °K

(Personal communication, 1985 and 1986, C.Warner, Department of Meteorological Research, Oxford). This model does not handle the multiple-scattering through the ice clouds.

Then, the measurement fields in the different AMSU channels (TB) are obtained by location of the centers of the AMSU FOVs on the AVHRR orbit, and by application of the antenna pattern of the corresponding channel to the microwave brightness temperature field of the corresponding frequency computed in the previous step.

The cloud contamination in every AMSU channel (DTB) is the difference between the AMSU measurement (TB), computed in the previous step, and the clear brightness temperature (TBcl)

3.2 THE PRODUCT OF THE SIMULATION

Twenty maps of contamination by clouds of the AMSU channels can be computed for each AVHRR orbit. Two examples of those maps (AMSUA 89 Ghz and AMSUB 89 Ghz) are shown in Fig. 5 and Fig. 6, the corresponding AVHRR infra-red channel 4 is presented in Fig. 4 and the atlas map in Fig. 3.

4. RESULTS

4.1 PRINCIPLES OF THE CLOUD CONTAMINATION IN AMSU

4.1.1 Simple rules for a quick estimation of the contamination by clouds in the different channels:

The effects of a cloud on a microwave channel can be very quickly estimated as follows:
    -The height of the peak of the weighting function (defining the absorbing/emitting layer of the atmosphere at that frequency) indicates whether the cloud has an effect (if the cloud is above the peak) or not (if the cloud is below the peak).
    -The effect of an ice cloud is to reduce the brightness temperature (by scattering), especially for high frequencies.
    -The effect of a water cloud is to reduce the brightness temperature (by absorbtion) and to emit at the temperature of the cloud (TC). The resulting effect has the same sign as the contrast (TC-TBcl) where TBcl is the clear microwave brightness temperature, whose order of magnitude is shown in table 3. The effect is stronger when the contrast is important and when the absorption coefficient has a high value.

TABLE 3. CLEAR BRIGHTNESS TEMPERATURES AT THE AMSU FRE-
QUENCIES COMPUTED WITH THE U.S. STANDARD ATMOS-
PHERE.

```
atmosphere          :   U.S standard atmosphere
emissivity of land:     .95
emissivity of sea  :    computed with a wind-speed of 10 m/s
```

channel / frequencies(Ghz)	clear brightness temperature (degree)	
	land	sea
1 / 23.8 Ghz	270.6	155.4
2 / 31.4 Ghz	269.9	152.4
3 / 50.3 Ghz	269.4	218.8
4 / 52.8 Ghz	262.7	251.7
5 / 53.6 Ghz	251.6	249.5
6 / 54.4 Ghz	237.0	236.9
7 / 54.9 Ghz	227.7	227.7
8 / 55.5 Ghz	221.4	221.4
9 / 57.3 Ghz	218.2	218.2
10 / 57.3 Ghz	220.1	220.1
11 / 57.3 Ghz	224.5	224.5
12 / 57.3 Ghz	231.6	231.6
13 / 57.3 Ghz	242.2	242.2
14 / 57.3 Ghz	254.0	254.0
15-16 / 89.0 Ghz	271.6	211.9
17 / 166.0 Ghz	275.3	262.8
18 / 183.3 Ghz	243.4	243.4
19 / 183.3 Ghz	256.7	256.7
20 / 183.3 Ghz	269.9	269.5

H. LEGLEAU

TABLE 4. EFFECT OF A CLOUD FILLING THE ENTIRE FIELD OF
VIEW OF THE AMSU ON THE DIFFERENT FREQUENCIES OF
THE AMSUA AND AMSUB.

```
atmosphere          :    U.S standard atmosphere
emissivity of land:      .95
emissivity of sea :      computed with a wind-speed of 10 m/s
ice cloud           :    top temperature: 240 degree - LWC: 200g/m2
water cloud         :    top temperature: 260/280 degree - LWC: 200g/m2
```

channel / frequencies(Ghz)	water cloud contamination (degree)		ice cloud contamination (degree)	
	land	sea	land	sea
	260deg/280deg	260deg/280deg		
1 / 23.8 Ghz	+0.1/+0.5	+8.5/+5.4	0.0	0.0
2 / 31.4 Ghz	+0.2/+0.8	+14.0/+9.4	-0.1	0.0
3 / 50.3 Ghz	-0.6/+1.0	+10.0/+9.1	-0.4	-0.3
4 / 52.8 Ghz	-1.0/ 0.0	+1.5/+2.1	-0.4	-0.4
5 / 53.6 Ghz	-0.7/ 0.0	-0.2/+0.4	-0.3	-0.3
6 / 54.4 Ghz	-0.2/ 0.0	-0.2/ 0.0	-0.2	-0.2
7 / 54.9 Ghz	-0.1/-0.1	0.0/ 0.0	-0.1	-0.1
8 / 55.5 Ghz	0.0/ 0.0	0.0/ 0.0	0.0	0.0
9 / 57.3 Ghz	0.0/ 0.0	0.0/ 0.0	0.0	0.0
10 / 57.3 Ghz	0.0/ 0.0	0.0/ 0.0	0.0	0.0
11 / 57.3 Ghz	0.0/ 0.0	0.0/ 0.0	0.0	0.0
12 / 57.3 Ghz	0.0/ 0.0	0.0/ 0.0	0.0	0.0
13 / 57.3 Ghz	0.0/ 0.0	0.0/ 0.0	0.0	0.0
14 / 57.3 Ghz	0.0/ 0.0	0.0/ 0.0	0.0	0.0
15-16 / 89.0 Ghz	-0.5/+3.1	+18.3/+23.8	-4.5	-2.5
17 / 166.0 Ghz	-3.2/+1.5	+2.8/+8.3	-38.6	-35.1
18 / 183.3 Ghz	-0.1/ 0.0	-0.1/ 0.0	-17.0	-17.0
19 / 183.3 Ghz	-0.8/ 0.0	-0.8/ 0.0	-34.6	-34.6
20 / 183.3 Ghz	-2.9/+0.2	-2.7/+0.2	-44.3	-44.2

TABLE 5. EXTREME VALUES OF THE CONTAMINATIONS   BY   CLOUDS
        (DTB)   OF   THE DIFFERENT   AMSU FREQUENCIES FOUND
        WITH OUR SIMULATION.

emissivity of land: .95
emissivity of sea :   computed with a wind-speed of 10 m/s

channel / frequencies(Ghz)   extreme brightness temperature changes
                                due to cloud contamination

channel / frequencies(Ghz)	land	sea
1 / 23.8 Ghz	+1.5	+2.0
2 / 31.4 Ghz	+3.0	+4.0
3 / 50.3 Ghz	+2.0/-2.0	+3.0
4 / 52.8 Ghz	-2.0	+0.5
5 / 53.6 Ghz	-1.0	+0.2
6 / 54.4 Ghz	-0.3	0.0
7 / 54.9 Ghz	0.0	0.0
8 / 55.5 Ghz	0.0	0.0
9 / 57.3 Ghz	0.0	0.0
10 / 57.3 Ghz	0.0	0.0
11 / 57.3 Ghz	0.0	0.0
12 / 57.3 Ghz	0.0	0.0
13 / 57.3 Ghz	0.0	0.0
14 / 57.3 Ghz	0.0	0.0
15 / 89.0 Ghz	+3.5/-6.3	+7.0
16 / 89.0 Ghz	+3.5/-10.0	+10.0/-1.5
17 / 166.0 Ghz	-35.0	+2.0/-10.0
18 / 183.3 Ghz	-25.0	-7.0
19 / 183.3 Ghz	-30.0	-8.0
20 / 183.3 Ghz	-30.0	-8.0

4.1.2 Order of magnitude of the cloud contaminations:

The effects of a cloud filling the whole field of
view can be estimated with table 4.

We can notice that the contamination by water clouds
at the AMSUA frequencies can reach high values over the
sea and that the ice clouds may have a very important effect on AMSUA 89 Ghz and AMSUB channels (one must keep in
mind that the results for an ice cloud are very dependent
on the ice particle size distribution).

4.2 CLOUD CONTAMINATION INFERRED FROM A REALISTIC
CLOUD FIELD

We studied three different situations, corresponding
to day orbits in April,1985, over mid-latitude regions.
All sorts of clouds were present: low broken clouds, cu-
mulonimbus and low widespread thick clouds over land, low
and high widespread clouds over ocean. The extreme values
of the cloud contaminations, as encountered under those
realistic conditions are shown in table 5. We can notice
that:
-the temperature sounding channels (channels 5-14 of AM-
SUA) are nearly unaffected by clouds.
-the water vapor content sounding channels (channels 17-
20 of AMSUB) are very much affected by ice clouds.
-all the channels are less affected by clouds than could
be expected from the values given in table 4; this must
be due to the size of the FOV, which is large enough to
be only rarely completely overcast.

4.3 QUANTITATIVE STUDY OF THE CONTAMINATION BY CLOUDS
OF THE AMSUA WINDOW CHANNELS

4.3.1 Over sea:

The cloud contaminations in the four AMSUA window
channels 1, 2, 3 and 15 (DTB1, DTB2, DTB3 and DTB15 respectively) over sea are not independant, as shown on
Fig. 7, Fig. 8 and Fig. 9. Those figures show that it
should be possible to regress the cloud contamination in
channels 2, 3, and 15 from that in channel 1, irrespective of the scanning angle, the cloud type and the vertical
structure of the atmosphere.

Those regressions are possible because the ice
clouds have no effect on channels 1, 2, and 3, and apparently little effect on channel 15, because the sea has a
low emmissivity and because the channels 1, 2, 3, and 15
are rather good window channels. Then, the ratio of the
cloud contaminations in two different window channels (n)
and (m) is given in Eq.1, where the term, in the external

brackets, is nearly independent on cloud temperature, sea emissivity and atmospheric profile and is nearly equal to 1.

$$\frac{DTB(n)}{DTB(m)} = \frac{ka(n)}{ka(m)} * (1 + \frac{Dmn(TBcl) - TC*Dmn(ta*ta*(1-es))}{(TC-TBcl) + TC*ta(m)*ta(m)*(1-es(m))}) \quad (1)$$

where
  ka is the absoption coefficient of the cloud
  es is the emissivity of the sea
  ta is the transmittance of the atmosphere
  Dmn(x) = x(m) - x(n)

   We can therefore create three new "clear channels":
     TB21 = 1.7 * TB1 - TB2
     TB22 = TB3 - 0.7 * TB2
     TB23 = 2.9 * TB3 - TB15
which are not affected by clouds over sea and which then depend only on the characteristics of the clear atmosphere, the sea surface emissivity, and the rain.

4.3.2 Over land:

   The previous results are not true over land, as it can be seen on Fig. 7, Fig. 8 and Fig. 9; this is due to the high emissivity of the land (assumed to be equal to 0.95).

4.3 QUANTITATIVE STUDY OF THE CONTAMINATION BY CLOUDS OF THE AMSUB WINDOW CHANNELS

   No similar results can be obtained for the AMSUB window channels, as it can be seen on Fig. 10. This is due to the great importance of the scattering process in the ice clouds at those frequencies.

         5. POSSIBLE USE OF THE PREVIOUS RESULTS

5.1 DETECTION OF PRECIPITATION OVER SEA IN THE AMSUA FOOTPRINT

   The brightness temperatures in "channels" 21, 22 and 23 depend only on the characteristics of the clear atmosphere, the sea emissivity and the precipitations. Then, it should be possible to regress the brightness temperature in channel 23 (TB23) from that in channels 21 and 22 ... if there is no precipitations, as they depend only on the transmittance of the clear atmosphere (depending essentially on water vapor) and on the sea emissivity (a known function of frequency and wind speed). The diffe-

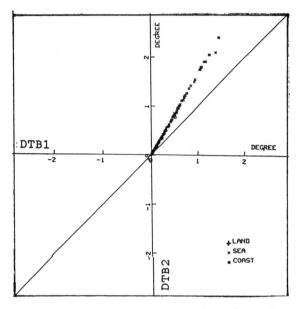

Figure 7. Comparison of the contamination by clouds in
          channel 1 (AMSUA/23 Ghz):DTB1 against that in
          channel 2 (AMSUA/31 Ghz):DTB2.

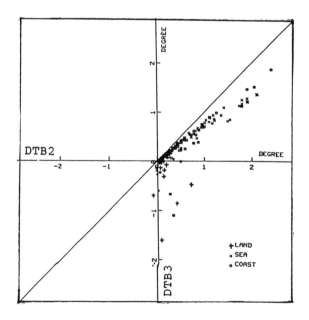

Figure 8. Comparison of the contamination by clouds in
          channel 2 (AMSUA/31 Ghz):DTB2 against that in
          channel 3 (AMSUA/50 Ghz):DTB3.

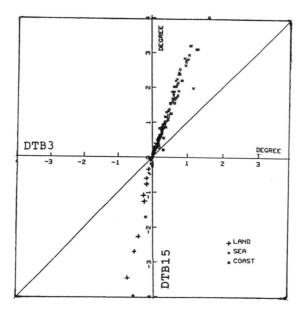

Figure 9. Comparison of the contamination by clouds in channel 3 (AMSUA/50 Ghz):DTB3 against that in channel 15 (AMSUA/89 Ghz):DTB15.

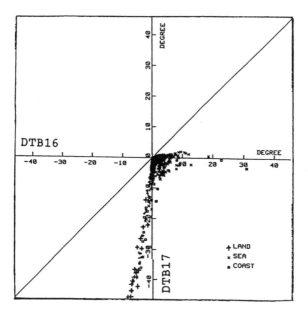

Figure 10. Comparison of the contamination by clouds in channel 16 (AMSUB/89 Ghz):DTB16 against that in channel 17 (AMSUB/166 Ghz):DTB17.

rence between the real TB23 and the regressed TB23 should then be a good indicator of precipitation in the AMSUA footprint.

5.2 DECLOUDING OF THE AMSUA WINDOW CHANNELS OVER THE SEA

When no rain has been detected, it should then be possible to get the clear values of the window channels 1, 2, 3 and 15 by using a regression between the clear brightness temperature in those channels and the three regressions between the cloud contaminations (computed previously).

6. CONCLUSION

It has been shown that the brightness temperature of the AMSU window channels will be affected by nonprecipitating clouds. The main interest of this simulation is to show the possibility to detect rain in the AMSUA FOV over sea and to retrieve clear AMSUA channels over sea if no rain is detected. It would be interesting to study the effect of polarization of the AMSU channels on the validity of the regressions between the cloud contaminations in the AMSUA window channels.

ACKNOWLEDGMENTS

I would like to thank C. Warner, R.W. Saunders, K.T. Kriebel and J. Eyre for their guidance through this work done in the Department of Meteorological Research at Clarendon Laboratory, Oxford, during 1985-1986.

REFERENCES

Calhon, D.,1984: Performance and Operation Specifications for the Advanced Microwave Sounding Unit.

Kriebel, K.T., 1986: Optical Properties of Clouds from AVHRR/2 Data. in American Meteorological Society (Eds),Sixth Conference on Atmospheric Radiation, 81-84.

Pick, D., 1985: Specifications of the Advanced Microwave Sounding Unit B.

Pruppacher, H.R.,Klett, J.D., 1978: Microphysic of Clouds and Precipitation , Reidel,D (Eds).

Saunders, R.W., 1986: Retrieval of Surface and Clouds Parameters from AVHRR/2 data'. in American Meteorological Society (Eds), Sixth Conference on Atmospheric Radiation, 78-80.

Saunders, R.W.,Kriebel, K.T., 1988: An Improved Method for Detecting Clear Sky and Cloudy Radiances from AVHRR data , Int.J.of Remote Sensing, 9 123-151

Warner, C., 1984: Satellite Observation of a Monsoon Depression , Final Report to NASA under Grant NAG 5-297.

THE INCORPORATION OF INVERSION CHARACTERISTICS INTO GROUND-BASED MICROWAVE TEMPERATURE SOUNDINGS: A SIMULATION STUDY

Patricia A. Miller
NOAA/ERL/PROFS
Boulder, Colorado 80303, USA
and
Michael J. Falls
NOAA/ERL/WPL
Boulder, Colorado 80303, USA

Results of radiometer temperature profile simulations are analyzed in order to examine the hypothesis that knowledge of temperature inversion parameters obtained from other instruments would substantially improve the accuracy of radiometric temperature profiles. Five different temperature retrieval algorithms are presented and compared with radiosonde data under both inversion and non-inversion conditions. The best algorithm yields consistently better results than the traditional (pure radiometric) technique, but still fails to perfectly reproduce the radiosonde inversions.

1. INTRODUCTION

Passive microwave radiometers that measure atmospheric radiances from the ground have been used and studied at NOAA's Wave Propagation Laboratory (WPL) in Boulder, Colorado for nearly two decades (Hogg et al., 1983). They hold particular promise for mesoscale meteorology because of their ability to derive a number of atmospheric parameters at a rate that can be considered nearly continuous in time. The accuracies associated with these products are, however, quite variable. For instance, accuracies for integrated quantities such as total precipitable water vapor and pressure layer thicknesses are at least competitive with, in some cases better than, radiosonde accuracies for the same parameters (Hogg et al., 1983). In contrast, temperature profiles derived from radiometric data are often considered deficient by meteorologists because they lack detail.

Particularly troublesome is the radiometer's inability to recover the sharp details of meteorologically significant temperature inversions. This characteristic can be attributed to limited vertical resolution (caused by broad radiative transfer weighting functions) and has led to the hypothesis that the addition of inversion information from other instruments would significantly improve radiometric temperature retrieval (Westwater,1978; Gossard and Frisch, 1987). Work in this area has already shown promising results. For example, Westwater et al. (1983) showed that inferred tropopause parameters are more accurate if the height of the tropopause is known. Here, we discuss simulations we performed with profiles that include middle-

MICROWAVE REMOTE SENSING
of the EARTH SYSTEM
Alain Chedin (Ed.)

and low-level inversions. In particular, information from temperature
inversions (height, depth, and strength) is evaluated for improving
the radiometric temperature retrieval up to 5 km above the surface.

A significant aspect of these simulations is the derivation of
error statistics conditioned on the presence of inversions. RMS
(root-mean-square) differences of radiometer and radiosonde temperatures of 2 K are often reported for fixed heights in the
atmosphere. However, these RMS errors do not reveal the radiometer's
accuracy in the important event of an inversion. Our study isolates
radiometer performance under conditions of inversions embedded in the
troposphere.

The radiosonde data used for this study consist of nearly 3000
soundings taken during the Atmospheric Variability Experiment (AVE)
Program under the sponsorship of the Marshall Space Flight Center,
NASA (Hill et al., 1979). The area covered by the soundings spans
roughly from the Great Plains to the Atlantic coast. Most of the
observations were obtained at 3-h intervals, the rest at 1.5- or 6-h
intervals. Each AVE observation period was characterized by frontal
zones, strong convective activity, or subsidence inversions somewhere
in the region covered by the observations. The routine rawinsonde
data were obtained by NWS upper-air stations at 12-h intervals. The
AVE data set was thus attractive because we could expect frequent
observations of inversion and non-inversion cases that occurred.
Hence, it seems reasonable to assume that this data set provides a
realistic basis for a preliminary error analysis. A summary of the
characteristics of the inversions observed in this data set is given
in Section 3.

Radiometric profiles corresponding to the AVE profiles were simulated using the method described below. Simulated profiles were
necessary because the AVE program did not include ground-based
microwave radiometers. In fact, large ground-based radiometric data
sets that incorporate the geographic and meteorological variability of
the AVE data set do not exist. However, experience with both ground
and satellite-based microwave radiometers (Westwater et al., 1985) has
shown that microwave radiances can be calculated to within about 0.5
K, so the profiles simulated from the radiances can be considered
accurate.

2. METHOD OF ANALYSIS

Temperature profile retrieval simulations were carried out on the
AVE data set as follows: 1) Radiances were calculated for each AVE
profile, at each of the six frequencies of the currently existing WPL
radiometer, by using standard absorption and radiative transfer models
(Waters, 1976); 2) all radiances were combined with the observed temperature inversion parameters to obtain equations for five different
retrieval methods; 3) Gaussian random noise with a standard deviation
of 0.5 K to represent radiometer noise was added to the radiances; and
4) the retrieval equations were applied to the simulated radiometer
data set to obtain simulated temperature profiles that could be compared with the observed temperature profiles.

Equations for the five retrieval methods were derived using various extensions of linear statistical estimation (Strand and Westwater, 1968). This procedure first calculates coefficients $[c_i(h), i=0,\ldots,9]$ from radiosonde soundings and then combines them with a nine-component data vector $[d_i, i=1,\ldots,9]$ to yield the temperature estimate $T(h)$ at height h.

$$\hat{T}(h) = C_o(h) + \sum_{i=1}^{9} C_i(h)d_i \tag{1}$$

The data vector, in this case, consisted of four brightness temperatures at 52.85, 53.85, 55.45, and 58.8 GHz; two optical depths (derived from brightness temperatures at 20.6 and 31.65 GHz); and surface observations of pressure, temperature, and relative humidity.

Coefficents for the retrieval equations were derived in two ways. First, the entire set of AVE soundings was used in the calculation. These coefficients, in combination with the data vector given above, produce the traditional (pure radiometric) equation used in our first retrieval method. Coefficients for our remaining retrieval methods, however, were calculated from each of several subsets of the AVE profiles, which are formed by grouping those profiles containing inversions based within selected height intervals. The decision as to which set of coefficients to use in obtaining any particular simulated profile was based on the inversion height measurement obtained from the corresponding radiosonde profile. For example, if we had a radiosonde profile with an inversion based at height h, we used coefficients that were calculated from the ensemble of profiles, whose inversion bases included those at height h, to derive the simulated radiometer profile.

Combining the coefficients derived from the various sets of radiosonde profiles with the nine-component data vector described in equation 1) gives the equations for our second retrieval method. This particular technique, first developed by Westwater and Grody (1980), was the method used in deriving the tropopause inversion parameters in the Westwater et al. (1983) study mentioned above. Several features of this method are notable. First, an independent inversion height measurement is used but only as a decision variable; the actual value of the inversion height does not enter into the formulation. The accuracy of the inversion location, therefore, is dependent only upon the resolution of the stratification layers used in generating the retrieval coefficients. The stratification layers for this study are listed in Table 1 (set 2). They were chosen as a compromise between sample size and resolution. Notice that, on the basis of the resolution of these layers, the best retrievals will be those profiles with inversions based from 0-1 km above the surface, and the worst will be those with inversions between 3-5 km above the surface. Above 5 km, the sample size (as is seen in section 3) did not allow for stable coefficient generation. The second notable feature of the Westwater and Grody (1980) method is the one-to-one correspondence between retrieval equations and observed inversions. This correspondence implies that, because each retrieval equation defines an entire tem-

TABLE 1. PROPERTIES OF THE FIVE RETRIEVAL TECHNIQUES USED IN ANALYSIS

Set	Parameters
1	All profiles (radiance) regardless of existence of inversions were used.
2	Profiles (radiance) were stratified into subsets before retrieval equations were derived. Each subset was defined by the existence of a temperature inversion base within a predefined height interval. Height intervals were 0-0.1, 0.1-0.4, 0.4-0.7, 0.7-1.0, 1.0-1.5, 1.5-2.0, 2.0-2.5, 2.5-3.0, 3.0-5.0, and 5.0 km. Each subset of retrieval equations was applied to the derived radiances on the basis of the layer within which the inversion was observed.
3	Same as Set 2, with the addition of the actual height of the inversion base within the subset, as a parameter in deriving the subsets of retrieval equations.
4	Same as Set 3, with the addition of the depth of the inversion layer having its base within the subset layer, as a parameter in deriving the subsets of retrieval equations.
5	Same as Set 4, with the addition of the temperature change through the inversion layer having its base within the subset layer, as a parameter in deriving the subsets of retrieval equations.

perature profile, multiple radiometer profiles are produced for each radiosonde profile with multiple inversions. This fact made comparing radiometer and radiosonde temperature profiles difficult. We chose to solve the problem by combining the multiple radiometer profiles in such a way that the structure around inversions would be preserved. We used the following algorithm: if $\hat{T}_c$ is the combined retrieval resulting from m separate retrievals T_i^c then

$$\hat{T}_c(h) = \frac{\sum\limits_{i=1}^{m} W_i(h)\hat{T}_i(h)}{\sum\limits_{i=1}^{m} W_i(h)} \qquad (2)$$

where

$$W_i = \exp[-(h-h_i)/2(wt)**2]$$

$$h = \text{height above the surface}$$

h_i = height (above the surface) of the ith inversion

m = the number of inversions

wt = weight given the inversion

In this study, wt was empirically set to 50.

Retrieval methods 3-5 were devised by including the inversion height; the inversion height and depth; and the inversion height, depth, and strength, respectively, to the data vector described in equation (1). The coefficients used were again computed using the stratification procedure described above. As in method 2, the inversion information was taken directly from the AVE radiosonde data and thus represented "perfect measurements" of the parameters. If inclusion of these inversion parameters results in significant improvement in the accuracy of the derived profiles, one has a rationale for directing attention to remote means of sensing these inversion parameters. Some achievements in this area have already been reported (Gossard and Frisch, 1987).

In summary, a radiometric profile of temperature is derived using equation (1) and specific information from the structure of each inversion. If the profile has no inversions, all five retrieval methods will give the same results. If inversions are present, each method will produce a different profile. The properties of our five sets of retrieval equations are briefly stated in Table 1. Separate error analyses, presented in Section 4, were performed on the temperature profiles obtained from each retrieval technique. In the next section, we present a statistical summary of the temperature inversions observed in the AVE data set.

3.   DATA SET DESCRIPTION

In this section we present a sample climatology of the inversions observed in the AVE data set. The AVE data are available at 25 mb intervals or at every contact point in the rawinsonde ascent (3-5 mb intervals). We used the data obtained at every contact point. A total of 2898 soundings were analyzed to produce the inversion sample climatology presented here. Inversions were arbitrarily defined by the requirement that $\Delta T \geqslant 2°C$ and $\Delta P \geqslant 10$ mb; they totaled 8932.

The distribution of inversions by height intervals above ground is presented in Fig. 1. Nearly one-third of the soundings contained inversions with bases between the surface and 1 km. Roughly one-seventh contained inversions with bases between 1 and 2 km. The number of inversions fell off rapidly between 2 and 8 km; above 8 km the number increased again as tropopause inversions began to be observed. The histogram in Fig. 1 illustrates a significant aspect of low- and middle-level tropospheric inversions. Except for surface-based radiation inversions (with lifting bases as the morning solar

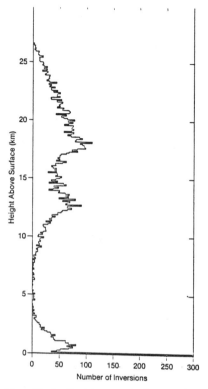

FIGURE 1.   The distributions of inversions by
            height intervals for the AVE data set.
            Inversions are defined by the conditions
            $\Delta T \geqslant 2°C$ and $\Delta P \geqslant 10$ mb.

heating occurs), all other tropospheric inversions can be termed statistically rare events.   That is, frontal surfaces or subsidence
inversions, the two major inversion phenomena other than radiation or
tropopause inversions, rarely occur within a given height interval.
This behavior has immediate implications for deriving retrieval
equations or computing temperature profiles.   If one derives retrieval
equations using the entire sample of any given data set, the data are
heavily biased toward the non-inversion cases.   Furthermore, the computed temperature profiles will have RMS errors that are only slightly
affected, if at all, by the occurrence of the rare or occasional temperature inversion.   Conversely, if one wants to derive retrieval
equations based only on profiles containing inversions, one should use
a data sample a thousand times (or more) larger than the AVE data
sample to obtain statistically robust results.   Considering that
rawinsonde data are routinely available only in early morning and
early evening hours, obtaining a proper-sized data sample is not a
trivial task.   Recall from Table 1 that the sets of profiles to be
analyzed are based on retrieval equations derived from subsets of data

stratified by inversions occurring in various layers between the surface and 5.0 km. The sample climatology summarized in Fig. 1
strongly suggests that these derived equations will be statistically
less reliable for the subsets from, say, 2 km up through 5 km.

   Another aspect of the inversion sample climatology is presented
in Figs. 2-6. These show the distributions of inversion strength ΔT,
inversion depth ΔZ, and inversion gradient $\Delta T/\Delta Z$, for the layers 0-1,
1-2, 2-3, 3-4, and 4-5 km above the surface. We note that the distributions of ΔT and ΔZ are highly skewed toward the right; i.e., strong
inversions or deep inversions occur relatively rarely. Yet these are
the cases that are probably most meteorologically significant. The
distributions of $\Delta T/\Delta Z$ are also skewed to the right with a modal value
near $10^{-2}\,°C\ m^{-1}$. Note that the ordinates are labeled differently in
some of these figures.

   Figure 7 contains histograms for ΔT, ΔZ, and $\Delta T/\Delta Z$ for all the
inversions in the data set based between the surface and 5 km. These
histograms underscore the skewness of the distributions as well as the
modal value of $\Delta T/\Delta Z$. Although it may be risky to do so, it seems
worthwhile to speculate about the universality of these distributions. If they are indeed drawn from a representative sample, they
suggest that the occurrence of temperature inversions within a specified height interval is a rare event. Further, relatively strong or
deep inversions are even more rare. This suggests that the specifi-
cation of temperature inversions from radiosonde-based retrieval
equations will be an extremely difficult task. In that context, it
is well to consider the present work as a pilot study for analysis
based on more robust statistics that might be developed in the future.
In particular, it should be noted that in this study the retrieval
equations are applied to the same set of radiances that they were derived from (with the exception that Gaussian noise was added). This
technique is not optimal, but it is common in cases such as ours where
the sample size is simply not large enough to support both a development and a validation set of profiles.

4.   ERROR ANALYSIS

   The RMS differences between the simulated radiometer profiles and
the observed radiosonde profiles are shown in Fig. 8. These error
plots include a non-inversion case, where points not contained in an
inversion were used exclusively in calculating the RMS differences, as
well as five inversion cases, where only inversion points were used.
The inversion cases used the five different algorithms described in
Table 1. Only one non-inversion case is needed because the five
algorithms differ only in their treatment of inversions; non-inversion
statistics will, therefore, be the same for all five methods. The
error bars on the inversion points represent 95% confidence intervals.
The fact that their length increases with increased height is due to
the decreasing number of inversions at greater heights. Error bars
are not included on the non-inversion RMS differences because the
sample sizes were so large that the 95% confidence intervals are all
less than 0.05°C. Notice that the radiosonde/radiometer differences

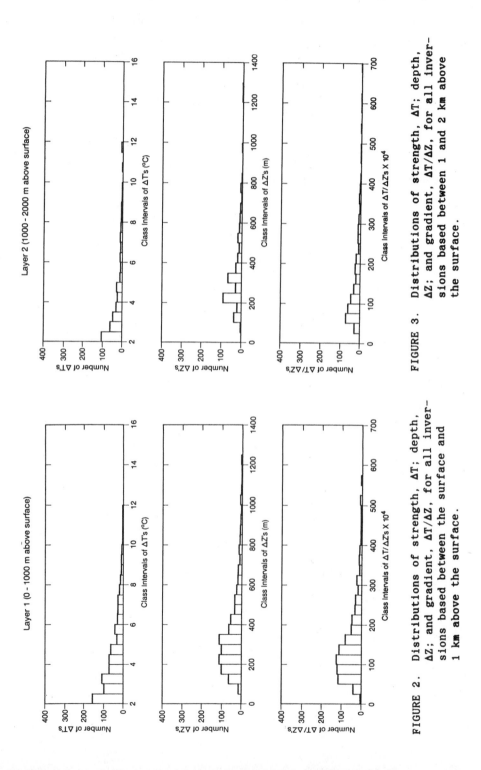

FIGURE 2. Distributions of strength, ΔT; depth, ΔZ; and gradient, ΔT/ΔZ, for all inversions based between the surface and 1 km above the surface.

FIGURE 3. Distributions of strength, ΔT; depth, ΔZ; and gradient, ΔT/ΔZ, for all inversions based between 1 and 2 km above the surface.

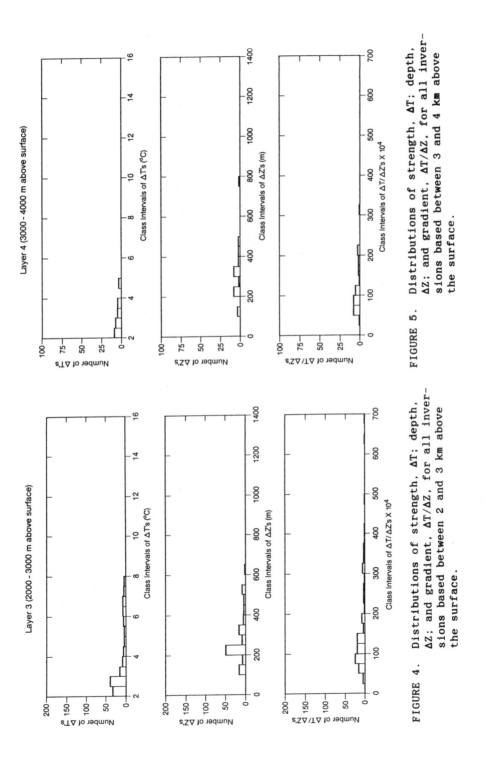

FIGURE 5. Distributions of strength, ΔT; depth, ΔZ; and gradient, ΔT/ΔZ, for all inversions based between 3 and 4 km above the surface.

FIGURE 4. Distributions of strength, ΔT; depth, ΔZ; and gradient, ΔT/ΔZ, for all inversions based between 2 and 3 km above the surface.

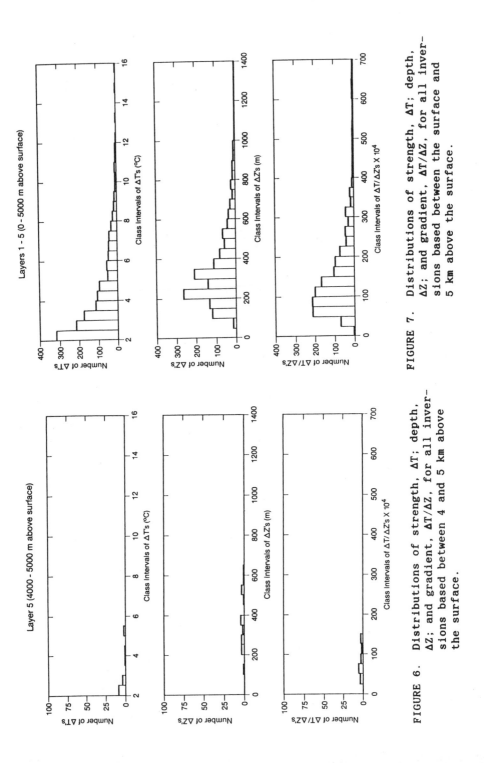

FIGURE 7. Distributions of strength, ΔT; depth,
ΔZ; and gradient, ΔT/ΔZ, for all inversions based between the surface and
5 km above the surface.

FIGURE 6. Distributions of strength, ΔT; depth,
ΔZ; and gradient, ΔT/ΔZ, for all inversions based between 4 and 5 km above
the surface.

were not calculated at particular heights through the atmosphere, but instead they were calculated for five layers. These atmospheric layers are 0-1, 1-2, 2-3, 3-4, and 4-5 km above the surface.

The following conclusions can be drawn from Fig. 8 concerning RMS errors for inversion points: First, we see a general trend of decreasing RMS error with increasing method number which is statistically significant in the lower layers. In other words, there is some evidence that the accuracy of the retrieval technique increases with the addition of more and more inversion parameters. The second conclusion concerns the relative importance of the various parameters in that improvement. Notice that in the 0-1 km layer (where our statistics are best and our error bars the smallest) 65% of the total error reduction occurs between method 1 and method 2. This implies that the reduction achieved by adding the approximate inversion height (method 2) is twice as large as the reduction achieved by including the exact inversion height, depth, and strength (method 5). The explanation for this result probably lies, not in what information was used, but in how the information was used. Recall that, in method 2, the profiles used to generate the retrieval equations were first stratified into subsets based on the presence of inversions between specified height intervals. Each of these subsets was then used to generate a retrieval equation. Method 2 uses the input inversion height information to do nothing more than choose among the retrieval equations; it essentially forces an inversion based at the (approximate) input height into every profile it produces. If we had further stratified the "generating" profiles into subsets based on inversion strength, the addition of ΔT could have had a significant effect (and we would be forcing an inversion of a specified depth into the retrieved profile at a specified height). In short, the a priori stratification has the most effect. Improvement is also seen from method 1 to method 2 for the layers above 0-1 km, although the percentage of the total reduction accounted for by the stratification is reduced from 65% to 50%, a difference easily accounted for by the lower resolution present in the stratification layers (see Table 1).

Figure 8 also shows significant differences between the non-inversion RMS errors and the inversion RMS errors. These differences exist through the 3-4 km level and are independent of the retrieval algorithm used. In other words, even with the exact height, depth, and strength specified for all inversions, none of our five methods produced RMS errors for inversion points as low as those errors produced for non-inversion points. The inversion RMS errors, however, are not dramatically larger than the non-inversion errors. This is due to the fact that the majority of inversions, in the AVE data set, are quite weak. Statistically, then, the expectation would be that RMS errors measured at inversion points would be relatively small even if the radiometric profiles were inversion free.

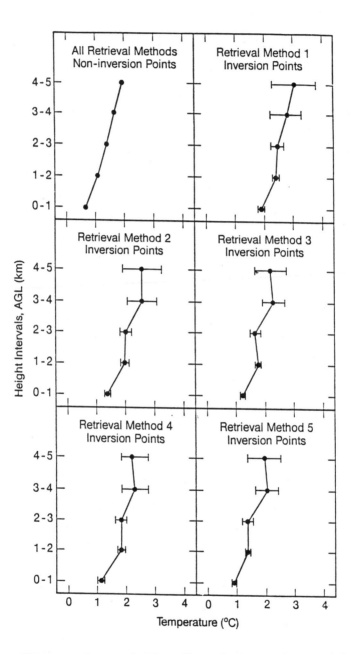

FIGURE 8.   RMS comparisons of AVE radiosonde temperatures with
            radiometric temperatures derived from simulated radiances
            using five retrieval methods. The five methods differ
            only in their treatment of inversion points.

Comparisons of inversion strengths derived from radiosonde and radiometric profiles are more revealing. Figures 9-14 show scatter plots of such comparisons. The inversion strengths were calculated from the AVE radiosondes; the radiometric temperature differences were calculated at the pressure heights corresponding to the radiosonde inversions. Recall from Section 3 that only (radiosonde) inversions with strengths of ≥2°C were considered in our data set. Realize also that points with negative abscissa values indicate cases where the radiometer profile is showing a temperature decrease in the same area where the radiosonde is showing an increase. Cases in which inversions are perfectly reproduced are represented by points lying on the diagonal line.

Figure 9 displays all inversions between the surface and 5 km above the surface on scatter plots for each of the five retrieval methods. These plots make it very clear that the a priori stratification is the major contributor to improving the structure of the radiometric temperature profiles. They also show that the improvement is substantial: in nearly all the cases the existence of an inversion is indicated. However, the scatter plots clearly show that all five retrieval algorithms produce inversions that are too weak. They also indicate that adding height, depth, and strength to the data vector does have a positive effect, although not a large one, on the performance of the retrieval technique. The exception is that the addition of ΔZ actually seems to make the retrievals worse. This is probably because inversions are a function of both height and temperature; adding ΔZ without ΔT is, therefore, not adding new information (but only noise) to the retrieval equations.

Figures 10-14 show the scatter plots for inversions based in layers 0-1,1-2 2-3, 3-4, and 4-5 km above the surface for the five retrieval methods. Also included for each method is a scatter plot for layer 0-1 km where all surface inversions were omitted. This latter plot makes it clear that ground-based radiometers are quite good at retrieving surface inversions, without the need to add inversion parameters. (The fact that the inversion is surface-based means that both the temperature and height at the inversion base are already included). Separating the scatter plots into layers also shows the degradation in performance with increased height that takes place despite the retrieval method used. This degradation is due not only to the fact that the radiometer loses resolution with height, but also to the fact that the stratification of inversions in retrieval method 2 also loses resolution with height. The result is shown in the scatter plots by the large percentage of points with negative abscissa values in the layers above 3 km.; i.e., most inversions are not reproduced correctly. The negative abscissa values do not necessarily represent profiles without an inversion but rather profiles having the inversion that was put in at the wrong height. In fact, for retrieval methods 2 through 5, we know that an inversion exists because the retrieval equations force an inversion into the profile. These cases could be substantially improved by a finer height stratification of the generating profiles above 3 km, something that could be accomplished only with a larger a priori data set of inversions.

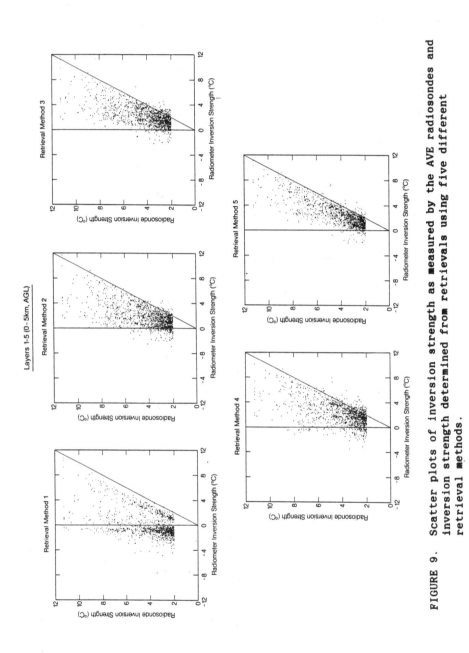

FIGURE 9.   Scatter plots of inversion strength as measured by the AVE radiosondes and inversion strength determined from retrievals using five different retrieval methods.

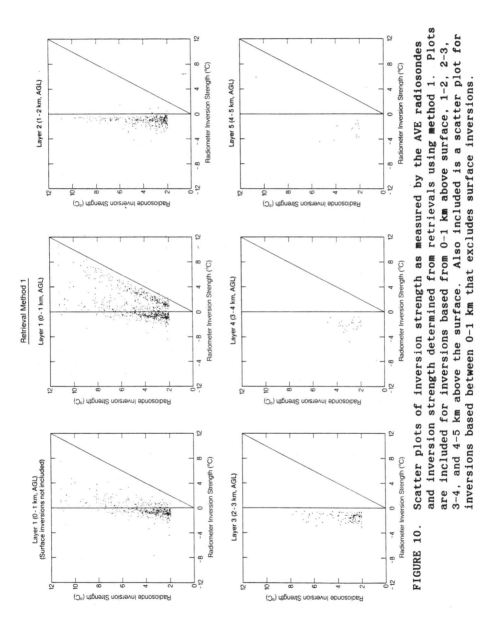

FIGURE 10. Scatter plots of inversion strength as measured by the AVE radiosondes and inversion strength determined from retrievals using method 1. Plots are included for inversions based from 0-1 km above surface, 1-2, 2-3, 3-4, and 4-5 km above the surface. Also included is a scatter plot for inversions based between 0-1 km that excludes surface inversions.

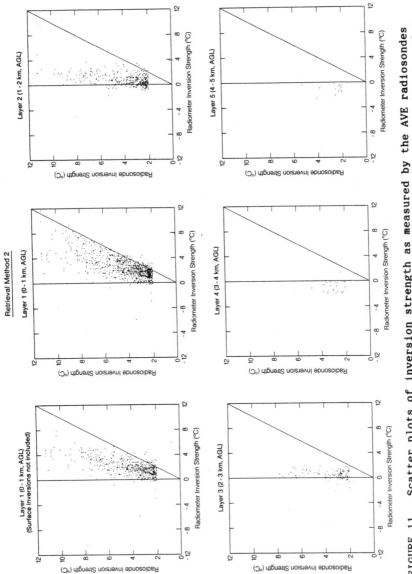

FIGURE 11.   Scatter plots of inversion strength as measured by the AVE radiosondes and inversion strength determined from retrievals using method 2. Plots are stratified into layers as in Fig. 10.

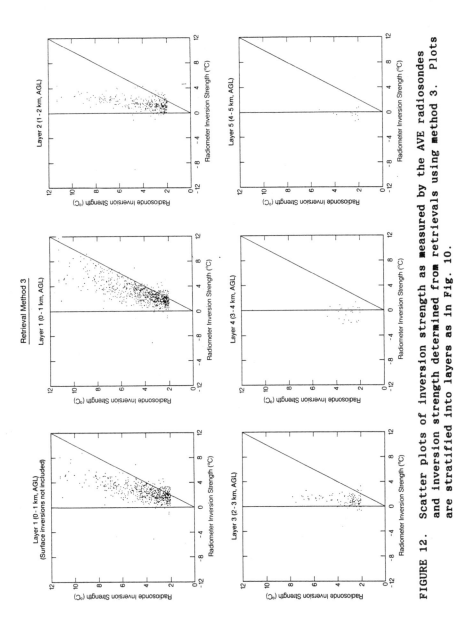

FIGURE 12. Scatter plots of inversion strength as measured by the AVE radiosondes and inversion strength determined from retrievals using method 3. Plots are stratified into layers as in Fig. 10.

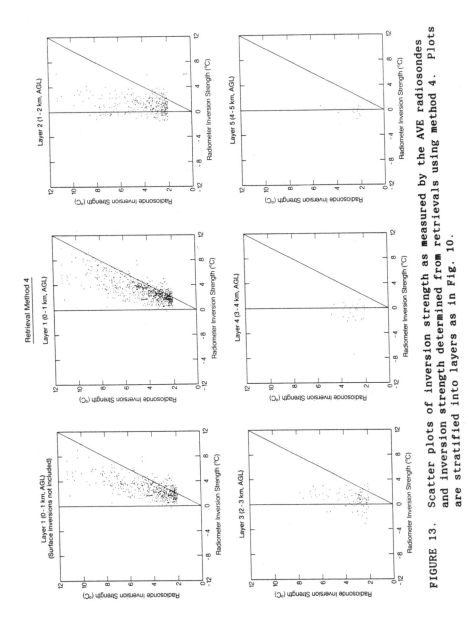

FIGURE 13.   Scatter plots of inversion strength as measured by the AVE radiosondes and inversion strength determined from retrievals using method 4. Plots are stratified into layers as in Fig. 10.

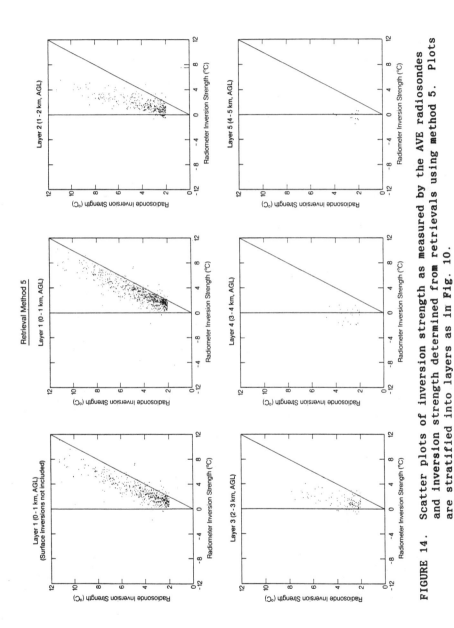

FIGURE 14. Scatter plots of inversion strength as measured by the AVE radiosondes and inversion strength determined from retrievals using method 5. Plots are stratified into layers as in Fig. 10.

     The final technique selected for comparing radiosonde and
radiometer temperature profiles is to plot five radiometric profiles
(and a single radiosonde profile) on the same axes. Figures 15-17 are
examples of such plots. The solid curve represents the temperature
measured by the radiosonde; the dashed curves are radiometric
retrievals calculated by methods 1-5. These plots best illustrate the
improvement that occurs when the augmented retrieval equations
(methods 2-5) are used. Notice, for instance in Fig. 16, how much
better the radiometer profile retrieved with method 5 matches the
radiosonde profile than the radiometer profile retrieved with method
1. These plots indicate that the role of the inversion parameters in
the retrieval equations for methods 3-5 is to fine tune the location,
depth, and strength of the inversion forced into the profile by the a
priori stratification of method 2. For example, the upper inversion
introduced by method 2 shown in Fig. 16 is located too high. Adding
the exact height of the inversion in retrieval method 3 shifts it down
to the correct height. Another example can be seen in Fig. 17. Here,
the addition of ΔZ and ΔT greatly improved the retrieval of a very
strong inversion. It is in extreme cases, where the inversion is not
typical of those used in generating the retrieval equations, that the
fine-tuning parameters add the most to the retrievals. Remember, no
attempt was made to reproduce inversions higher than 5 km above the
surface.

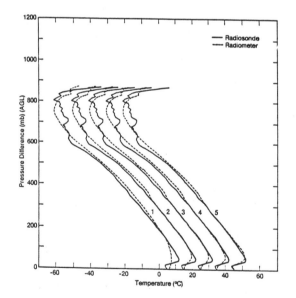

FIGURE 15.   Comparisons of a radiosonde temperature profile (solid
             curves) with radiometric profiles retrieved using five
             different methods (dashed curves). Profiles 2-5 are
             displaced to the right. All methods attempted to repro-
             duce temperature inversions existing in the radiosonde
             profile from the surface to 5 km above the surface.

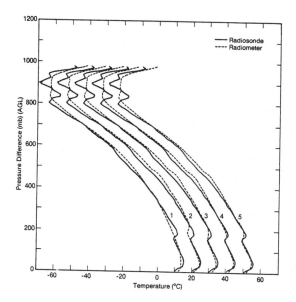

FIGURE 16. Similar to Fig. 15, except a different temperature profile.

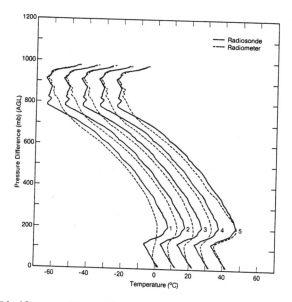

FIGURE 17. Similar to Fig. 15, except a different temperature profile.

5. CONCLUSIONS

The goal of this study is the examination of the hypothesis that the addition of inversion parameters from other sources would substantially improve the accuracy of radiometric temperature profiles in the vicinity of inversions. The results of the study are somewhat mixed. Inversion retrieval was definitely improved. With only the addition of inversion height, the retrieval methods correctly showed temperature increase where the traditional method (with no additional information) showed temperature decrease. However, even the addition of "perfect" measurements of inversion height, depth, and strength was not enough to guarantee excellent retrievals. Instead, the retrieved profiles showed a substantial bias toward inversions weaker than those present in the radiosonde profiles. In addition, none of the augmented retrieval methods produced RMS errors in inversion situations as small as those for non-inversion cases. Perhaps a retrieval method that stratified the generating profiles into subsets based on both inversion height and inversion depth would correct the bias and reduce the RMS error. Certainly, the a priori stratification method tested in this study shows more promise than those methods that simply added parameters to the retrieval equations. Unfortunately, we were unable to stratify on both height and depth, because such a stratification would require a much larger set of profiles than the 2898 that we had available. In fact, if our data set is representative, the set of profiles required (i.e., a multivariate distribution of both heights and strengths) would be very difficult to collect.

Our study used data representing a wide geographical area encompassing large variations in surface pressure. In fact, variations in surface pressure in the AVE data range over roughly 200 mb, a feature accounted for by using the surface pressure as a predictive parameter. Earlier studies investigating the accuracy in ground-based radiometry focused on performance at a single station. Figure 8 shows that our non-inversion RMS errors are nearly identical to the RMS errors calculated by comparing radiometer and radiosonde temperatures at one site (Decker, 1983). This finding would seem to indicate that surface pressure is an extremely valuable parameter in the retrieval process. If this result is substantiated by additional analyses, it could have far-reaching effects on the implementation of a system of radiometers. Currently, it is thought that retrieval techniques are so dependent on climatology that each radiometer would need its own set. These results imply that that may not be the case. That is, radiometers may not need to be at sites of radiosonde stations. This, in turn, would imply that radiometers have the capability to increase not only the temporal resolution of temperature profiles, but also the spatial resolution.

ACKNOWLEDGMENTS

The authors wish to express thanks to Duane A. Haugen and Ed R. Westwater for their constant encouragement and interest during this work. The authors also thank Paul Kent for preparing many of the computer programs.

REFERENCES

Decker, M.T., 1983: A comparison of radiosonde- and radiometrically-derived atmospheric observations. Proc. WMO/AMS/CMOS Fifth Symposium on Meteorology Observations and Instrumentation, April 11-15, Amer. Meteor. Soc., Boston, 205-206.

Gossard, E.E., and A.S. Frisch, 1987: The relationship of the variances of temperature and velocity to atmospheric static stability--Application to radar and acoustic sounding. Accepted by J. Climate Appl. Meteor.

Hill, K., G.S. Wilson, and R.E. Turner, 1979: NASA's participation in the AVE-SESAME 1979 program. Bull. Amer. Meteor. Soc., 60, 1323-1329.

Hogg, D.C., M.T. Decker, F.O. Guiraud, K.B. Earnshaw, D.A. Merritt, K.P. Moran, W.B. Sweezy, R.G. Strauch, E.R. Westwater, and C.G. Little, 1983: An automatic profiler of the temperature, wind, and humidity in the troposphere. J. Climate Appl. Meteor., 22, 807-831.

Strand, O.N., and E.R. Westwater, 1968: Minimum rms estimation of the numerical solution of a Fredholm integral equation of the first kind. SIAM J. Numer. Anal., 5, 287-295.

Waters, J.W., 1976: Absorption and emisssion by atmospheric gases, Chapter 2.3 in Methods of Experimental Physics, Vol. 12-Part B-Radio Telescope, Academic Press, New York, 142-176.

Westwater, E.R., 1978: Improved determination of vertical temperature profiles of the atmosphere by a combination of radiometric and active ground-based remote sensors. Preprints Fourth Symp. on Meteorological Observation and Instrumentation, Denver, Amer. Meteor. Soc., Boston, 153-157.

Westwater, E.R., and N.C. Grody, 1980: Combined surface- and satellite-based microwave temperature profile retrieval. J. Appl. Meteor., 19, 1438-1444.

Westwater, E.R., M.T. Decker, A. Zachs, and K.S. Gage, 1983: Ground-based remote sensing of temperature profiles by a combination of microwave radiometry and radar. J. Climate Appl. Meteor., 22, 126-133.

Westwater, E.R., Wang Zhenhui, N.C. Grody, and L.M. McMillin, 1985: Remote sensing of temperature profiles from a combination of observations from the satellite-based microwave sounding unit and the ground-based profiler. J. Atmos. Oceanic Tech., 2, 97-109.

THE INFLUENCE OF THE VERTICAL STRUCTURE OF HUMIDITY ON THE RETRIEVAL OF TOTAL WATER VAPOUR CONTENT OVER THE OCEANS BY MICROWAVE RADIOMETRY FROM SPACE

C. Simmer, E. Ruprecht, and D. Wagner
Institut für Meereskunde
D-2300 Kiel, FRG

ABSTRACT

A qualitative analysis of the radiative transfer equation is presented to understand how the vertical structure of the water vapour profile influences the radiation temperature at the top of the atmosphere. Radiative transfer calculations are carried out using a power-law vertical distribution of water vapour and two profile models including an atmospheric boundary layer. The calculations suggest changes of up to 8 K in the radiation temperature at the top of the atmosphere, and that these changes are caused by different vertical distributions but identical total amount of water vapour. These changes are dominant in the center and wings of the weak water-vapour absorption line (22.235 GHz) and almost zero at frequencies close to the center of the line. A study with a set of 400 measured high-resolution atmospheric profiles over the Atlantic Ocean confirms these findings. The RMS error due to the natural variability of water vapour profiles is found to be in the range of 0.5 kg m^{-2}.

1. INTRODUCTION

The retrieval of total water vapour content (W) with microwave radiometry from satellites is generally based on the physics stated below :

1) Water vapour is a dominant but weak absorber. This is true around the weak water-vapour absorption line at 22.235 GHz and in the window regions between the strong absorption lines of water vapour and oxygen (see Fig.1).

2) The emissivity of the earth surface is low, thus the surface contributes only little to the radiation temperature at satellite level. This is true for water surfaces for most satellite viewing angles where the emissivity is about 0.5.

3) No clouds and rain are present.

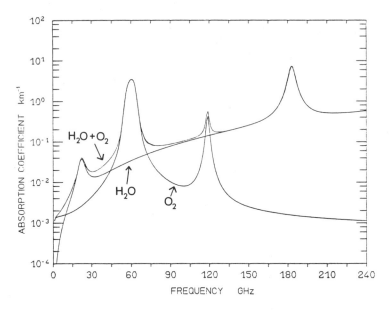

Figure 1. Absorption coefficient of O_2, H_2O, and O_2 plus H_2O for pressure of 1013.3 hPa, temperature of 288.15 K, and water vapour density of 7.5 g m^{-3}.

     With these assumptions the radiation temperature at satellite altitude is primarily a function of W alone. The variations of surface emissivity due to wind and foam, of sea surface temperature (SST), and the variability of the vertical profiles of water vapour and temperature are secondary effects. An analysis of the radiative transfer equation in this respect can be found in Grody (1976).

     Based on these assumptions a whole generation of passive microwave radiometers for satellites were designed and used accordingly with the frequencies near the weak water vapour absorption line at 22.235 GHz. Applications can be found for example in Grody (1976) and Chang and Wilheit (1979) for the NEMS instrument on NIMBUS-5, in Grody et al. (1980) for the SCAMS instrument on NIMUS-6, in Chang et al. (1984) for the SMMR instrument on NIMBUS-7, in Alishouse (1983) for the SMMR instrument on SEASAT, and in Pathak (1987) for the SAMIR instrument on BHASKARA II. Currently two satellites with passive microwave instruments in the mentioned frequency range are in orbit: the Japanese satellite MOS-1 and the US satellite DMSP with the SMM/I instrument. But the next generation of microwave sensors will use the window regions in the higher frequency ranges to accomplish better spatial resolution, e.g., MMS on the next generation of METEOSAT satellites and AMSU on the NOAA satellites.

In this paper we will concentrate on the effect of the highly variable vertical profile of water vapour in the troposphere on the radiation temperature T_B at satellite level. Although this is only a secondary effect compared with the total water vapour content, we will show that the effects can be large enough to cause significant errors and biases in the retrieval of W when global regressions between W and T_B are used .

2. UNDERSTANDING THE MECHANISM

First we shall qualitatively discuss the influence of the water vapour profiles on T_B. For that we concentrate on the region in the vicinity of the weak water-vapour absorption line at 22.235 GHz.

The direct contribution of the atmospheric water vapour to T_B is given by the following equation:

$$T'_B = \int_0^\infty k_a(z) \, \mathcal{S}_{H_2O}(z) \, T(z) \, e^{-\tau(z)} \, dz \qquad (1)$$

with k_a : mass absorption coefficient in m^2/kg
$\mathcal{S}_{H_2O}$: water vapour density in $kg \, m^{-3}$
τ : transmissivity of atmosphere from level z to top,

There are two main influences of the water vapour profile on T_B: temperature effect and absorption coefficient effect. An increase in both quantities leads to an increase in T_B, which will be discussed in the following.

2.1 TEMPERATURE EFFECT

The contribution of a certain atmospheric layer to T_B is proportional to its temperature. Because this contribution is weighted by the water vapour density in the layer, a redistribution of the water vapour will have the following effect: If the profile is changed in a way that water vapour is moved from lower to higher levels i.e., towards lower temperatures, then T_B is decreased according to equation (1).

2.2 ABSORPTION COEFFICIENT EFFECT

Since the absorption coefficient is dependent on pressure and temperature, two effects have to be considered:

First, the absorption coefficient decreases slightly
with rising temperature, mainly in the wings. In and
around the line center (see Fig. 2) there are, however,
only very small changes. If water vapour is moved from
lower to higher levels – from higher to lower temperatures – T_B is increased in the wings only. This obviously
reduces or overcompensates the temperature effect in the
wings described in 2.1.

Second, the absorption coefficient increases in the
line center and decreases in the wings when the pressure
is reduced (Fig. 3). If water vapour is moved from lower
to higher levels k_a will decrease in the wings and increase in the center of the line. As a consequence T_B
will be reduced in the wings and increased in the center.

The effect of pressure broadening is the strongest
one, as shown in Fig. 4, where all effects on the absorption coefficient are combined for a standard atmospheric
profile. Large effects of the water vapour profile can be
expected in the center of the line at 22.235 GHz and in
the far wings. The smallest effects are found around 20
and 25 GHz, where all curves intersect (profile-independent frequencies). These results are similar to those described by Hogg et al. (1983), who were looking for optimal suited frequencies for ground-based instruments.

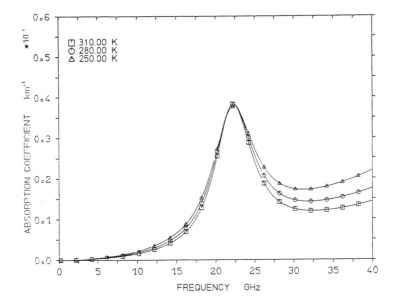

Figure 2. H_2O-absorption coefficient shown for 3 tempera-
        tures. Air-pressure is 1013.3 hPa and water
        vapour density is 7.5 g m^{-3}.

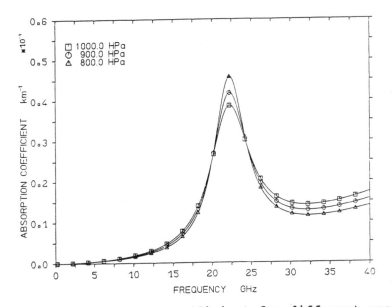

Figure 3. H_2O absorption coefficient for different pressures. Temperature is 280 K and water vapour density is 7.5 g m^{-3}.

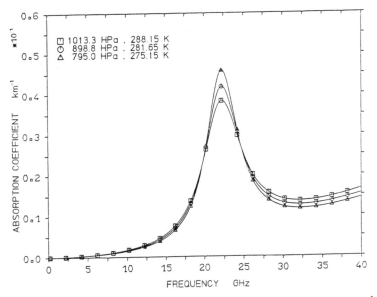

Figure 4. H_2O absorption coefficient for different air pressure and air temperature according to the US-Standard Atmosphere. Water vapour density is 7.5 g m^{-3}.

In this very qualitative discussion, we did neglect
the changes in the transmissivity of the atmospheric
layers above the layer under consideration due to the
redistribution of the water vapour. Thus, the exact location of the 'profile-independent' frequencies may still
shift to either side. In the next chapter the integral
effect will be investigated; therefore, artificial water
vapour profiles are constructed.

3. SIMULATION WITH MODEL PROFILES

In order to study quantitatively the effect of water
vapour profiles on the radiation temperature at satellite
level, we calculated the radiation transfer for artificial
atmospheric profiles. We used the formalism described in
Francis et al. (1983). The absorption coefficients of the
atmospheric constituents are taken from Ulaby et al.
(1981). Scattering is neglected in the cloudless atmosphere. The lower boundary is assumed to be a plane ocean
surface with salinity of 34 per thousand. The emissivity
and reflectivity are computed using the Fresnel equations.
The standard profile of the tropics is chosen for the
vertical temperature distribution, and the SST is equal to
the temperature of the lowest layer of the atmosphere.

The humidity profiles are described by a power-law
structure:

$$q(z) = q(0) \left[\frac{p(z)}{p(0)} \right]^{l} \quad , \quad l = const \qquad (2)$$

with   q : specific humidity
       p : pressure

As a reference profile we chose $l = 3.0$, which is
about the average for a large ensemble of measured profiles over the Atlantic Ocean. For individual profiles out
of this large ensemble we found values for l ranging
from 0.5 to 8.0. Fig. 5 shows such profiles as given by
equation (2) for l between 1.0 and 5.5 with constant
total water vapour content ($W = 45$ kg m^{-2}).

With these model profiles, radiative transfer calculations are carried out for 12 frequencies in the range
of the weak water vapour absorption line. The frequencies
chosen are identical to existing or planned channels on
satellites. The results will be given as deviation of
total optical depth (Fig. 6a) and of radiation temperature
(Figs. 6b-d) from the reference case.

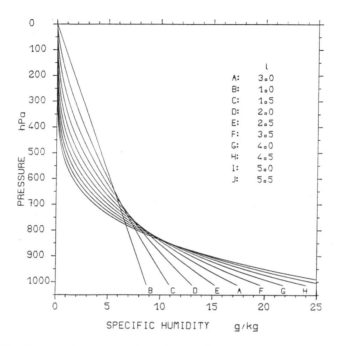

Figure 5. Power-law vertical profiles of specific humidity
with a constant total water vapour content of
45 kg m^{-2}

3.1 POWER-LAW PROFILES

If l is increased the profile structure is changed as
if water vapour were moved downwards to higher pressure
and higher temperature. This will decrease the optical
depth of the atmosphere in the center of the line as
expected and increase the optical depth in the wings of
the line as we assumed above. For a low emissivity background, this results in a lower T_B in the center of the
line and a higher T_B in the wings, as can be seen in Figs.
6b-d. They show T_B for nadir view (b), 50° view angle for
horizontal (c), and vertical polarization (d). The marked
differences between horizontal and vertical polarization
in the center of the line can be explained by the differences in the behaviour of the sea surface at higher
view angles. For a view angle of 50°, the emissivity of a
plane water surface is substantial higher for the vertical
than for the horizontal polarization. This adds a profile-
independent contribution of the sea surface to the signal
at the top of the atmosphere. Thus the effect of the ocean
surface relative to the effect of the atmosphere is increased. On the other hand, the reflectivity of the sea

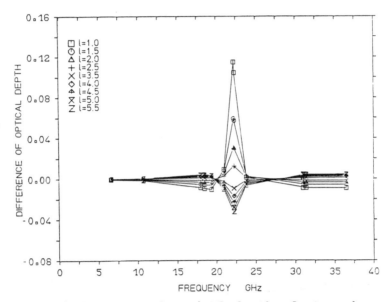

Figure 6a. Differences of optical depth of atmosphere due
to variation of power coefficient. Positive values indicate an increase of optical depth over
the profile with l=3.0.

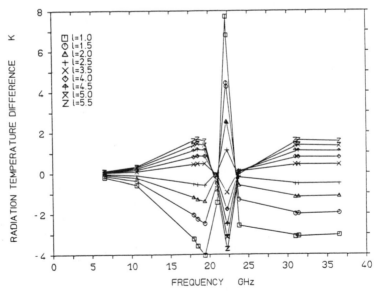

Figure 6b. Differences of radiation temperature at top of
the atmosphere due to variation of power coefficient. Positive values show increases of radiation temperature over the profile with
l=3.0. Calculations are made for nadir view.

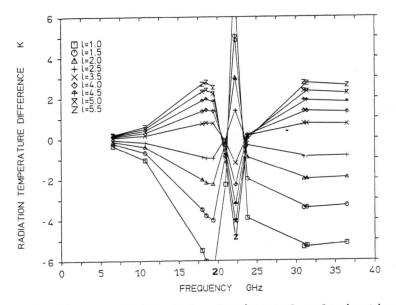

Figure 6c. Same as (b) but for horizontal polarization
           and view angle 50 deg.

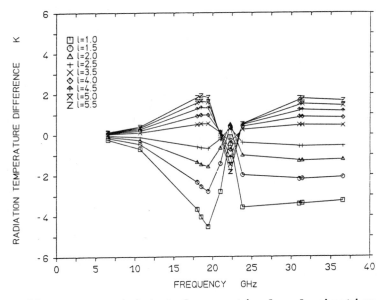

Figure 6d. Same as (c) but for vertical polarization   and
           view angle 50 deg.

surface is decreased. This reduces the part of the upward
reflected atmospheric radiation, causing a similar effect
as reducing the optical depth (emissivity) of the atmosphere.

3.2 BOUNDARY LAYER EFFECTS

The vertical structures of the water vapour discussed
in the previous section represent only one part of the
variability found in measured profiles. The build-up of an
atmospheric boundary layer (and also clouds) give rise to
large deviations from power-law profiles, causing, for
example, high concentrations of water vapour in lower
layers of the atmosphere. To test the effect of such
variations, we simulated the boundary layer by two models.
The total water vapour content is still held constant, and
the power-law profile with l=3.0 serves as reference.

MODEL 1:   We assume constant specific humidity within the
boundary layer. Its value is given by the surface value of
the reference profile. Above the boundary layer a power-
law profile is fitted by the assumption W = constant. Different profiles are constructed by changing the depth of
the boundary layer (Fig. 7a).

MODEL 2:   The specific humidity is again assumed constant
within the boundary layer. Its value, however, may vary.
Above the boundary layer the reference profile is assumed.
With increasing depth of the boundary layer and W = constant the surface value has to decrease (Fig. 7b).

We present only the results of the radiation transfer
calculations for nadir view angle. For MODEL 1 we have a
redistribution of the water vapour from upper layers of
the atmosphere into the boundary layer, which results in a
decrease of the radiation temperature in the center of the
line and an increase in the bands (Fig. 8a). The resulting
differences are comparable in magnitude to the results
from the previous sections for increasing power coefficient. Differences of up to 4 K can be expected.

For MODEL 2 the opposite behaviour is observed. An
increase of the boundary layer depth moves humidity up-
wards and leads to a radiation temperature increase in the
line center and a decrease in the wings (Fig. 8b). The
magnitude of the differences, however, is much less than
that of MODEL 1 (note different scales of the ordinates
in Fig. 8a and 8b). Assuming that both models represent
the natural variability of vertical profiles, we can draw
the conclusion that, in case of well-developed boundary
layers, the radiation temperature at top of the atmosphere
will be several degrees below average in the center of the
line and above average in the wings of the line.

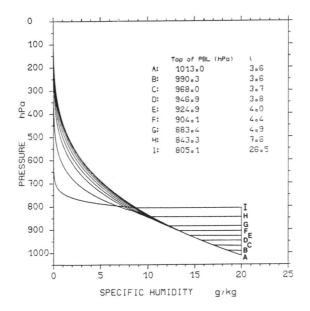

Figure 7a. Vertical profile of specific humidity for
different boundary layer depths with MODEL 1.

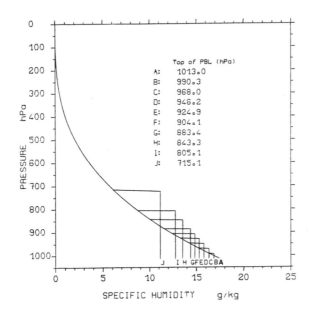

Figure 7b. Vertical profile of specific humidity for
different boundary layer depths with MODEL 2.

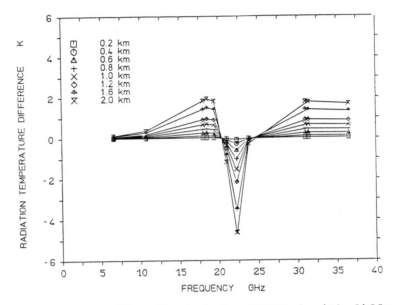

Figure 8a. Same as Fig. 6b, but for MODEL 1 with different
boundary layer depths.

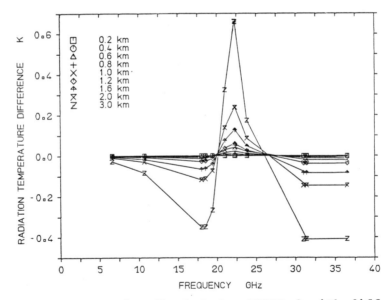

Figure 8b. Same as Fig. 6b, but for MODEL 2 with different
boundary layer depths.

For both models the vertical profile of temperature
(and pressure) is kept constant. A near-constant temperature profile in the boundary layer, as opposed to a near-
linear decrease, would possibly weaken the effect in the
center of the line and strengthen it in the bands.

4.   SIMULATIONS WITH MEASURED PROFILES

With the models of vertical water vapour profile
introduced in the previous sections, we intend to describe
the variability of the humidity in the troposphere. In
nature the moisture distribution is not as smooth as given
by these models. To study the effect of the natural variability, we compared radiative transfer calculations using
measured water vapour profiles with power-law profiles,
both having identical total water vapour content.

For the comparison, we used about 400 profile measurements from several expeditions of the research vessels
METEOR and POLARSTERN in the Atlantic Ocean during the
years 1982 - 1985. The radiosonde ascents were used with a
vertical resolution of 5 hPa from ship-level to 300 hPa.
Layers above 300 hPa were neglected for the study.

Assuming that no clouds are present, the radiation
temperature at the top of the atmosphere is mainly dependent on the total water vapour content W. We show this
relationship in form of scatterplots for frequencies
representative of the center of the line (22.235 GHz, see
Fig. 9a), of the off-center regions (21 GHz and 23.8 GHz,
see Figs. 9b and 9c), and of the wings (18 GHz and 31
GHz, see Fig. 9d). Total water vapour content is calculated for each radiosonde ascent, then the radiation
temperatures are derived with the radiative transfer
equation applying: a) the observed humidity profile, and
b) the power-law profile fitted to it with l = 3.0 and the
same water vapour content. For a qualitative comparison,
the two scatter diagrams are plotted with an offset of 10
K in the abscissa values.

The scatter in the diagrams for the observed profiles
is due to the variability of the vertical structure of
humidity and temperature, and of the sea surface temperature. For the power-law profiles, the variability of the
vertical humidity structure is subtracted.

Comparing the scatterplots for the three frequency
regions, we observe in accordance with the qualitative
findings of the previous chapter:

a) The center-of-line channel (22.235 GHz) and the
wing channels (18 and 31 GHz) show generally more scatter
about the regression curve than the off-center channels.

b) The wing channels show a marked reduction in scatter when the water vapour profile variability is subtracted (Fig. 9d). The reduced sensitivity for low water vapour contents in the 31-GHz channel is caused by the near O_2 - absorption band around 60 GHz (see Fig. 1). The O_2 absorption increases the temperature profile dependency of the radiation temperature, especially when the strong signal due to water vapour is low.

c) The center-of-line channel (22.235 GHz) shows a similar behaviour due to the subtraction of the water vapour profile variability. Beside the much higher W - sensitivity of this channnel (larger dT_B/dW) we observe also a marked nonlinear dependency in both cases for higher water vapour contents. The water vapour in the higher layers of the atmosphere screens the signal from lower layers resulting in a saturation of T_B.

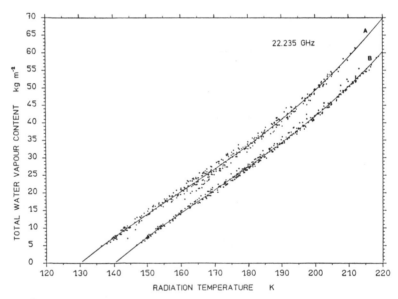

Figure 9a. Scatterplots of radiation temperature at top of the atmosphere (T_B) versus total water vapour content (W) from a set of 400 high-resolution radiosonde profiles over the Atlantic Ocean. T_Bs are computed by radiative transfer calculations with a plane ocean surface as lower boundary condition. View angle is 0 deg. Results are shown for the center-of-line channel (22.235 GHz) for both the measured water vapour profiles (A) and the power-law profiles (B) with l = 3.0 (off-set 10 K).

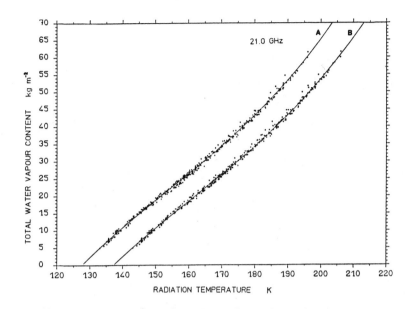

Figure 9b. Same as Fig. 9a, but for the close to center-
of-line channel 21 GHz.

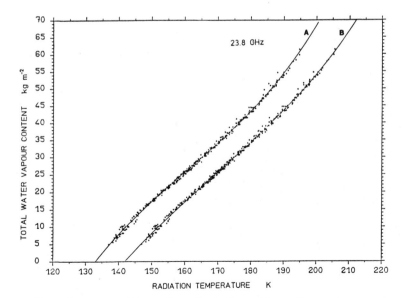

Figure 9c. Same as Fig. 9a, but for the close to center-
of-line channel 23.8 GHz.

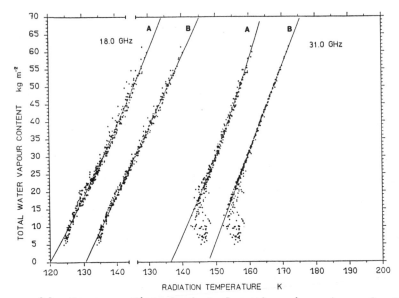

Figure 9d. Same as Fig. 9a but for the wing channels 18
and 31 GHz. Profiles with W < 15 kg m^{-2} were
excluded for the computation of the regression
line.

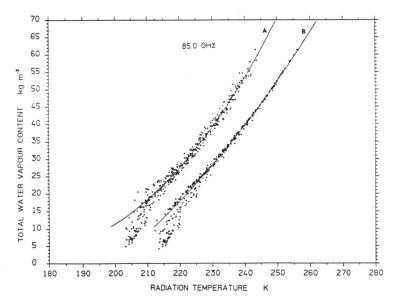

Figure 9e. Same as Fig. 9a but for 85.5 GHz. Profiles with
W < 15 kg m^{-2} were excluded for the computation
of the regression line.

d) The off-center channels show similar but reduced W – sensitivities compared with the center of line channel (Figs. 9b and 9c). There is no marked reduction in scatter when the variability due to the vertical structure of the water vapour profile is removed.

We performed the same calculations for one of the window-channels in the higher frequency range namely, for 85.5 GHz, which is part of the SMM/I radiometer on the DMSP satellite (Fig. 9e). We find similar results as for the other profile-dependent channels. The scatter is even more reduced when the profile structure variation is subtracted.

Similar results were obtained for other view angles in both polarizations. To quantify the qualitative impressions of the scatterplots, we fitted polynomials up to 3rd degree to the data and computed the standard error of the curves (Table 1). Results for 50° view angle, vertical (V,) and horizontal polarization (H) are included in the table. As expected, the standard error is substantially reduced by the use of a fixed-structure water vapour profile for all channels except the off-center channels. The error produced by the uncertainty of the water vapour profile is in the range of 0.5 kg m^{-2} on the average.

TABLE 1.   STANDARD ERRORS IN RESPECT OF REGRESSION
           CURVES (kg m^{-2})

GHz	Nadir View meas.	p.-l.	50° (V) meas.	p.-l.	50° (H) meas.	p.-l.
18.0	1.53	1.02	1.80	1.31	1.07	0.43
19.4	1.02	0.69	1.30	0.96	0.76	0.35
21.0	0.59	0.60	0.86	0.89	0.47	0.45
22.235	0.96	0.63	1.00	0.95	0.92	0.55
23.8	0.57	0.55	0.84	0.81	0.42	0.39
31.0	1.32[1]	1.09[1]	1.28[1]	0.54[1]	1.11[1]	0.65[1]
37.0	1.37[2]	1.20[2]	1.22[2]	0.51[2]	1.18[2]	0.81[2]
85.5	1.32[2]	0.86[2]	1.95	0.90	1.27	0.67

[1] : cases with W < 15 kg m^{-2} excluded from regression
[2] : cases with W < 20 kg m^{-2} excluded from regression
V : vertical polarization; H : horizontal polarization
meas.: measured profiles;   p.-l.: power-law profiles

4.   CONCLUSIONS

The radiation transfer calculations with the model water vapour profiles show a profile dependency of radiation temperatures for wing (e.g., 18 and 31 GHz) and cen-

ter-of-line channels (22.235 GHz) of several Kelvin. No
variations due to the structure of the water vapour profile are found for off-center channels (21 or 23.8 GHz).
The results using measured water vapour profiles show the
same tendencies. If the latter are compared with a smoo-
thed structure of the water vapour (1 = 3.0 power-law
profile), it is observed that the larger uncertainties in
regressions between total water vapour content and radiation temperature at the top of the atmosphere are, in
fact, due to the vertical structure variability of the
water vapour profiles.

We can conclude that caution has to be applied to
single-channel retrievals of total water vapour content in
regions of highly variable water vapour profile structure,
as in the tropics. The use of global relationships between
satellite-derived radiation temperatures and total water
vapour content will cause a bias in regions with water
vapour profile structures different from average (e.g.
power-law type) profiles. Such biases can be expected in
regions of persistent well-developed boundary layers like
the high-pressure cells in the subtropics. An analysis of
satellite-derived radiation temperatures in view of these
effects is necessary.

On the other hand, combinations of center-of-line
channel and off-center measurements can help to identify
certain water vapour structures. At least, information
about large deviations from average profiles e.g., the
existence of a well-developed boundary layer might be
extracted. In order to improve the humidity retrieval with
microwave radiometry, such a dual-channel observation is
recommended for satellite-borne radiometers planned in the
future.

REFERENCES

Alishouse,J.C.,1983: Total Precipital Water and Rainfall
Determinations From the SEASAT Scanning Microwave Radiometer, J. Geophys. Res., 88, 1929-1935

Chang,A.T.C., and T.T. Wilheit, 1979: Remote Sensing of
Atmospheric Water Vapour, Liquid Water, and Wind Speed at
the Ocean Surface by Passive Microwave Technique from the
NIMBUS 5 Satellite, Radio Science, 14, 793-802.

Chang,H.D., P.H. Hwang, T.T. Wilheit, A.T.C. Chang, D.H.
Stealin, and P.W. Rosenkranz, 1984: Monthly Distributions
of Precipitable Water from the NIMBUS 7 SMMR Data, J.
Geophys. Res., 89, 5328-5334.

Francis,C.R., D.P. Thomas, and P.L. Windsor, 1983: The Evaluation of SMMR Retrieval Algorithms. In T.D. Allan (Ed.), Satellite Microwave Remote Sensing, John Wiley and Sons, New York, 481-498.

Grody,N.C., 1976: Remote Sensing of Atmospheric Water Content from Satellites Using Microwave Radiometry, IEEE Trans. Ant. Prop., AP-24,155-162.

Grody,N.C., A. Gruber, and W.C. Shen, 1980: Atmospheric Water Content over the Tropical Pacific Derived from the NIMBUS-6 Scanning Microwave Spectrometer, J. Appl. Met., 19, 986-996.

Hogg,D.C., F.O. Guiraud, J.B. Snider, M.D. Decker, and E.R. Westwater, 1983: A Steerable Dual-Channel Microwave Radiometer for Measurement of Water Vapour and Liquid in the Troposphere, J. Appl. Met., 22,789-806.

Pathak,P.N., 1987: Empirical analysis of passive microwave observations from Bhaskara-II SAMIR and remote sensing of atmospheric water vapour and liquid water, J. Clim. Appl. Met., 26, 3-17.

Ulaby,F.T., R.K. Moore, and A.K. Fung, 1981: Microwave Remote Sensing, Aktive and Passive, Volume I: Microwave Remote Sensing Fundamentals and Radiometry, Addison-Wesley Publication Company, London

ERRORS IN SATELLITE RAINFALL ESTIMATION DUE TO NONUNIFORM FIELD OF VIEW OF SPACEBORNE MICROWAVE SENSORS

Long S. Chiu
Applied Research Corporation
Landover, MD 20785, USA

and

Gerald R. North
Climate System Research Program
Texas A & M University
College Station, TX 77843, USA

and

David A. Short
Goddard Space Flight Center
National Aeronautics and Space Administration
Greenbelt, MD 20771, USA

ABSTRACT

Nonuniform rainrates within a field of view (FOV) and a nonlinear rainrate- microwave temperature (R-T) relation lead to a bias in the estimation of areal average rainrate from spaceborne microwave measurements. This bias is estimated from rainrate data collected during the Global Atmospheric Research Program Atlantic Tropical Experiment (GATE) using a R-T relation which is derived from the model results of Wilheit et al. (1979). The percent bias is about 25% (30%) for a footprint size of 8 km and increases to about 40% (45%) for a footprint size of 40 km for phase I (II) of GATE. In the large FOV limit of 280-km footprint size, the bias is 48% (50%).

An approximate formula which takes account of the effect of spatial inhomogeneity within the FOV and a nonlinear R-T relation is derived. To first order, the two effects multiply to produce the bias. The bias formula is applied to rain field models. For a rain field model where the autocorrelation function is defined by an exponential, the dependence of the bias on the ratio of the FOV size and the e-folding scale is very similar to the dependence calculated from the GATE data. For a Poisson Process rain field model, the bias formula shows an inverse dependence on the probability of rain. In this context, the lower percent bias for phase I of GATE can be understood in terms of the higher probability of rain during the same period.

This work is supported in part by the National Aeronautics and Space Administration through grants NAG-5-869 and NAS5-30083.

MICROWAVE REMOTE SENSING
of the EARTH SYSTEM
Alain Chedin (Ed.)

95

1. INTRODUCTION

There is a growing awareness of the importance of accurate measurements of global precipitation to the advancement of our knowledge of the dynamics of the oceans and atmosphere. The lack of a surface-based global network points to satellite monitoring as the ultimate mode of observation (Austin and Geotis, 1980). Techniques available for this remote sensing problem (Arkin, 1979; Barrett and Martin, 1981; Atlas and Thiele, 1981) are based mostly on empirical relations between cloud properties and precipitation. The usefulness of these techniques is hampered by the lack of surface data for calibration.

For this reason, many have turned to microwave remote sensing methods for estimating rain, because the microwave radiation interacts directly with falling hydrometeors (Wilheit et al. 1977). Over the oceans, a clear indicator of rainrate is the microwave emission at frequencies below about 20 GHz. Figure 1 shows a schematic diagram (idealized here as an exponential) of the relationship between the brightness temperature and the rainrate for a uniformly covered field of view (FOV) for a given columnar height of rain. At the low rainrates, the brightness temperature increases with rainrate due to absorption and re-emission of rain drops and saturates at about 15-20 mm/hr. The brightness temperature decreases at the higher rainrates (not represented here) as the effect of scattering by hydrometeors takes a more dominant role. Recent calculations show that the effect of scattering may be more important than previously thought (Wu and Weinman, 1984; Kummerow, 1987). The relation states that, when the FOV is _filled_ with a _uniform_ rainrate R, there will be a corresponding microwave brightness temperature, T. We shall refer to the relation as an R-T relation. For example, at 19.3 GHz (the frequency of the Electrically Scanning Microwave Radiometer flown on NIMBUS 5, or ESMR-5), the range of variation is about 150 K, with instrument noise of a few degrees.

If all FOVs are filled with uniform rainrates, and in the low rainrate regime where there is a one-to-one correspondence between R and T, the inverse R-T relation can be used to obtain a rainrate for an observed brightness temperature. Although precipitation systems span a wide spectrum, much of the variability is contained in the the cumulus scale which is of the order of a few kilometers. Figure 2 shows an instantaneous radar echo pattern observed during the Global Atmospheric Research Program Atlantic Tropical Experiment (GATE) in the Inter-tropical Convergence Zone (ITCZ) region in the Atlantic during the summer of 1974. Overlain is a schematic of NIMBUS-7 SMMR (Scanning Multichannel Microwave Radiometer) 37 GHz channel FOVs. It can be seen that, at least in this swath, _no_ FOVs are uniformly filled.

Nonuniform rainrates within a FOV and a nonlinear R-T relation lead to a bias (systematic error) in the estimation of area average rainrate. Within a FOV, rainrates are distributed (Fig. 1). There is an associated distribution of temperature (through the R-T relation). Let [.] denote areal average over the FOV. The desired

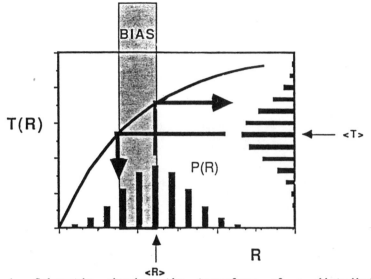

Figure 1. Schematic showing the transform of a distribution of rainrate, R --namely P(R)-- with mean ⟨R⟩, through a nonlinear function T(R), giving rise to a distribution in T, with mean ⟨T⟩. The difference between ⟨R⟩ and the rainrate estimated from ⟨T⟩ is called a bias.

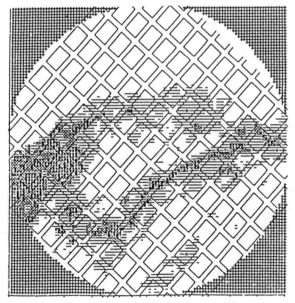

Figure 2. Schematic showing an instantaneous picture of radar echoes taken during GATE. Superposed on it are FOVs of SMMR footprints. Note that the FOVs are rarely uniformly filled.

quantity is [R] but only [T] is measured. Hence, in using the R-T relation, a rainrate of $R_E = R([T])$ is estimated. In general,

$$R([T]) \neq [R]$$

The error incurred in this inversion is

$$\delta R = [R] - R_E$$

A given [R] can give rise to many different R_Es because [T] depends on the distribution of R within the FOV.

The ensemble average of δR is called a bias. This bias, often referred to as the "beam filling" bias in remote sensing, has been studied by many investigators (Austin and Geotis, 1978; Smith and Kidder, 1978). It is, however, worth emphasizing that this bias is not due to unfilled FOVs alone. If the R-T relation is linear, as in the case of the retrieval of sea ice concentration from ESMR-5 (Zwally et al., 1983), there is no bias associated with the retrieval procedure.

The purpose of the paper is twofold. First, the dependence of the bias on resolution is examined using observed data from GATE. Qualitative knowledge of the dependence will be useful in the design of spaceborne precipitation experiments such as the proposed Tropical Rainfall Measuring Mission (TRMM) (Simpson et al., 1988). Second, simple models are proposed which delineate the effects of a nonlinear R-T relation and unfilled FOVs. It is hoped that experience gained from these models will provide rationale for the removal of the bias associated with rain estimation from microwave observations.

2.   BIAS ESTIMATES FROM GATE

2.1 GATE DATA

Radar-derived rainrate data, collected during GATE are used. GATE, was conducted in the summer of 1974. During roughly three triweekly periods, each termed "a phase," extensive radar and raingage measurements were made over an area called the B-scale area centered around 8.5°N and 23.5°W, encompassing an area of about 400 km in diameter. Arkell and Hudlow (1977) composited the radar measurements from research vessels and presented an atlas of the radar reflectivities every 15 minutes. Patterson et al. (1979) binned the data into 4-km by 4-km pixels and converted the radar reflectivity into rainrates. During phases I and II of GATE, the data were composited mainly from radar measurements collected by the C-band radar on board the research vessel "The Oceanographer," which was positioned at the center of the B-scale. During phase III, "The Oceanographer" was moved to the southeast quadrant. Only data from phases I and II are used in our study.

2.2 RAINRATE-TEMPERATURE RELATION

The R-T relation used in our study is a functional fit to the model results of Wilheit et al. (1977) for a rain column of 4 km and is of the form

$$T(K) = 274 - 102 \exp(-cR) \qquad R < 20 \text{ mm/hr}$$
$$T(K) = 276.44 - .22 R \qquad R \geq 20 \text{ mm/hr}$$

where T is the brightness temperature in K and R is rainrate in mm/hr. $c = 0.19$ (mm/hr)$^{-1}$. In this representation we assume that emission is dominant for rainrates less than 20 mm/hr and scattering takes over for higher rainrates.

2.3 BIAS CALCULATION

The original data are arranged in an array of 100 by 100 4-km pixels. Sixteen boxes, arranged in 4 by 4 arrays, are chosen within the whole array so that the left sides of the boxes is separated by 20 pixels (Fig. 3). This procedure attempts to minimize the effect of spatial dependence between neighboring boxes of the rainfall data.

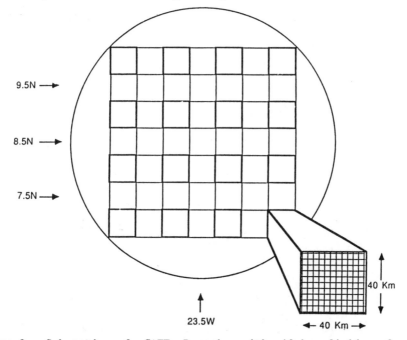

Figure 3. Schematic of GATE B-scale with 40-km fields of view superposed upon the 4-km bins. The shaded boxes indicate regions where the biases are calculated.

For different sizes of the boxes, the errors are calculated as

$$\delta R = [R] - R([T]) = [R] - R_E$$

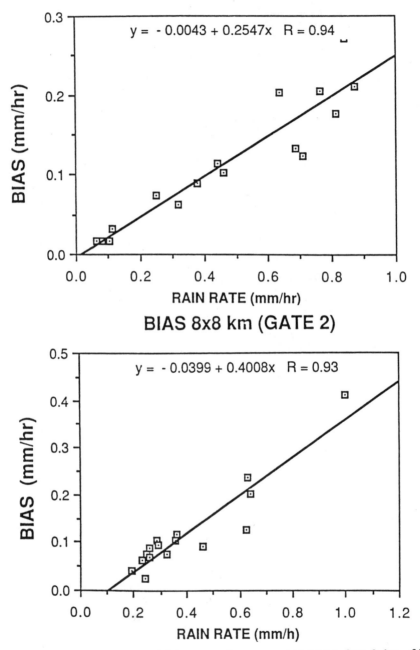

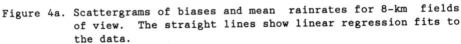

Figure 4a. Scattergrams of biases and mean rainrates for 8-km fields of view. The straight lines show linear regression fits to the data.

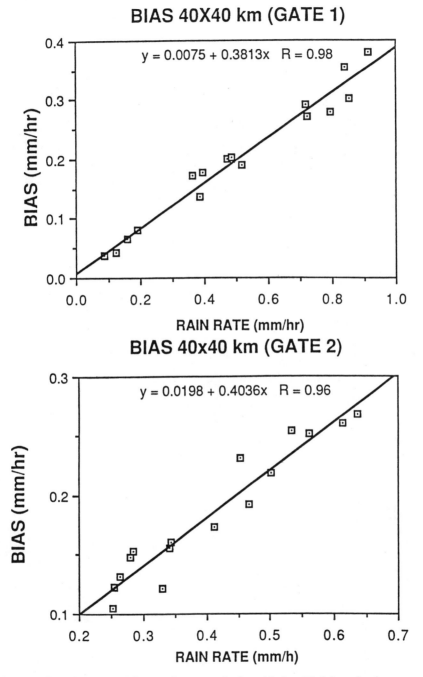

Figure 4b.   Same as Figure 4a except for 40-km fields of view.

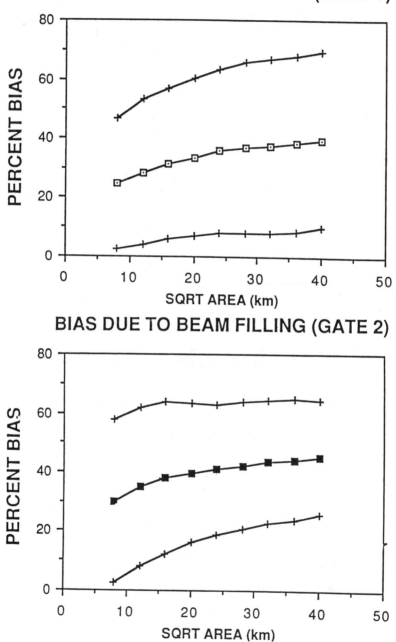

Figure 5. Calculated bias as a function of the size of FOVs for GATE I and II.

where again [.] denotes area average over the box whose size can be varied. The bias is obtained by ensemble averaging which is replaced by time averaging, denoted $<.>$. Biases as functions of the mean rainrates over boxes (FOVs) of side lengths of 8 and 40 km are depicted in Fig. 4. The biases are mostly linearly related to the mean rainrate in the FOV: linear regression analyses showed correlation coefficients larger than 0.9, which is significantly above the 99.5% level. Henceforth, the biases are expressed in terms of the percent of the FOV rainrate, $<\delta R> = \alpha<[R]>$. The true rainrate can be estimated from $R([T])$ as

$$<[R]> = R(<[T]>)/ (1-\alpha)$$

The percent bias is defined as

$$\beta = (<[R]> - R_E)/ <[R]>$$

The percent bias averaged over the 16 samples for different FOV sizes are presented in Fig. 5 for phases I and II. The error bars indicate one standard deviation calculated from the 16 samples. It can be seen that the percent bias increases from about 25% (30%) for FOV of 8 km to about 40% (45%) for FOV of 40 km for phase I (II). These curves suggest an asymptotic limit for large FOVs. The bias for a FOV of 280 km (i.e., taking all data in the large square as shown in Fig. 3) is calculated. The bias is 0.223 mm/hr (0.185 mm/hr) for phase I (II) which has an average rainrate of 0.464 mm/hr (0.364 mm/hr), hence a percent bias of 48.1% (50.7%).

3.   THEORETICAL CONSIDERATIONS

3.1 AN APPROXIMATE FORMULA

From section 1, it is seen that the bias depends not only on the mean, but also on higher moments of the rainrate distribution. In this section, an approximate formula for the bias is derived which can be considered a rule of thumb for estimating the bias.

Consider the function $T(R)$ of the R-T relation as shown in Fig. 1. For simplicity, we consider a one-to-one functional between R and T, i.e., no scattering. Let R_E be the estimated rainrate from the measured temperature [T], i.e.,

$$[T] = T(R_E) \tag{1}$$

A Taylor expansion of $T(R)$ about the averaged rainrate [R] gives

$$T(R) = T([R]) + (R-[R])T' + 1/2 (R-[R])^2 T'' + \ldots \tag{2}$$

where T' and T" are the first and second derivatives of T with respect to R, evaluated at [R]. If we evaluate equation (2) at $R=R_E$ and retain only linear terms, we obtain

$$T(R_E) \approx T([R]) + (R_E-[R])T' \tag{3}$$

If we apply the area averaging operator to (2), the linear term vanishes and we get

$$[T(R)] \approx T([R]) + 1/2 \ [(R-[R])^2 T"] \tag{4}$$

Recognizing that $T([R]) = T(R_E)$ from (1), and equating (3) to (4), an approximate expression for the error in estimating [R] from [T] is obtained

$$\delta R = [R] - R_E \approx -[(R-[R])^2] \ T"/2T' \tag{5}$$

Taking the ensemble average of (5) yields the bias. The formula (equation 5) is neatly factored into two parts. The first is a property of the rain field only, namely, the variance with respect to the averaging area (the mean square deviation from the areal mean over the FOV). The second is a function of the R-T relation only. This formula is consistent with our earlier notion that the error is due to nonuniform rainrate over the FOV and a nonlinear R-T relation. It is intuitively pleasing to see that the two effects multiply to produce the net result.

The sign of the bias can also be estimated to first order from the formula. The first factor (variance) is always positive and depends on the variability of rainrates in the rain field. The second term depends on the ratio of the curvature and the slope of the T(R) curve of the R-T relation. The term $T"/T'=-c$ is negative. Hence a positive bias is anticipated in the retrieval of rainrates from ESMR-5: R_E underestimates [R] in most cases. This is consistent with our results in section 2 and those of the previous investigators.

The formula indicates that the error is dependent on the rainrate variance within the FOV. This can be tested by using the GATE data. If we take the logarithm of both sides of equation (5), we get

$$\log \delta R \approx \log \ [(R-[R])^2] + \log T"/2T'$$

From the GATE data, the logarithm of the rainrate variance explains about 80% of the variance of the logarithm of the error. The histograms of log [R], log δR and log $[(R-[R])^2]$ for a 40-km FOV have been calculated (not shown). When plotted on a logarithmic scale they all show a bell-shaped distribution, suggesting that they are lognormally distributed.

We next apply equation (5) to some simple models of rain field and examine the associated biases.

3.2 POISSON PROCESS MODEL

Consider subdividing the FOV into N square tiles. Let x of the N tiles be raining with rainrate r_0 and in the rest of the N-x tiles, there is no rain. Let the probability of rain in an individual tile be p. If p is small and the probability of rain in one tile is independent of the other, we may adopt a Poisson model of the rain field. For a given realization of the process, the area average rainrate in the FOV is

and
$$[R] = x \ r_0/N$$
$$[R^2] = x \ r_0^2/N$$

Taking ensemble averages and using the properties of Poisson statistics, we find that the percent bias is

$$\beta = - r_0 \ (\ 1-p-1/N \ ) \ T''/2T' \tag{6}$$

From analysis of GATE data, the average rainrate conditional on positive rainrates for 4-km pixels is about 4 mm/hr and the p's are 12% and 9% for phase I and II respectively (Chiu, 1988). Since β depends on $(1-p)$, all other parameters being the same, the percent bias should be larger in phase II than in phase I since p is larger for phase I, which is consistent with Fig. 5. In the limit as $N \to \infty$ and $p \to 0$, which corresponds to the large FOV case, the percent bias becomes

$$\beta = r_0 \ c/2$$

if we use the same model of T(R) adopted in section 2. Putting in numerical values, β is about 40%.

3.3 RAIN FIELD WITH A LENGTH SCALE

The bias formula can be expressed as a high pass filter of the spatial spectrum. The case of a one-dimensional rain field is derived in the appendix. Consider the case where a length scale exists in the rain field, the autocorrelation function can be represented as

$$\rho \ (x) = \exp(-|x|/\lambda)$$

The corresponding spectral density function is

$$|R_k|^2 = |R_0|^2 \ 2/\lambda \ (1/\lambda^2 + k^2)^{-1}$$

If we substitute the spectrum into equation (A1), we get

$$\langle \delta R \rangle = T''/2T' \ R_0^2 \ I(\gamma)$$

where $I(\gamma)$ is the integral

$$I(\gamma) = 1/2\pi \int_{-\infty}^{\infty} 2\gamma/(\gamma^2+x'^2) \ [1-(\sin(x')/x')^2] \ dx'$$

and $\gamma=a/\lambda$ is the ratio of the size of FOV to the length scale of the rain field. Figure 6 shows the integral $I(\gamma)$ as a function of γ. The integral increases to a maximum of about 0.8 for $\gamma=20$ and decreases slowly as γ increases. As $\gamma \to \infty$, $I \to 0$. The ratio of $I(\gamma=4)$ to $I(\gamma=20)$ is about 0.75, which may be compared with the ratio of the biases for FOVs of 8 km to 40 km (Fig. 5). In the limit when $a \gg \lambda$, $I(\gamma) \to 0$. It seems that the formula breaks down in the large FOV limit in the one-dimensional case.

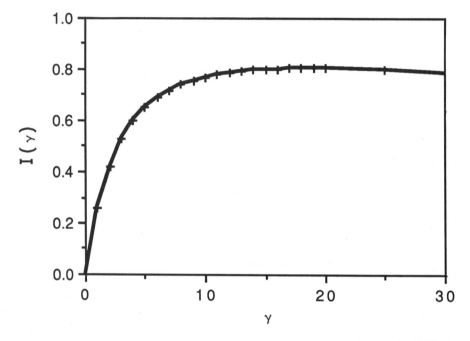

Figure 6. $I(\gamma)$ versus γ where γ is the ratio of the size of FOV to the length scale in the rain field.

4. SUMMARY AND DISCUSSION

The bias associated with nonuniformly filled FOVs of spaceborne microwave sensors has been estimated from radar data collected during GATE. An approximate formula is derived which shows that this bias is closely related to the variance of rainrate within the FOV of the sensor. By applying simple models of rain field to the formula, we show that formula is consistent with variation of the bias.

For the one dimensional case, our formula breaks down in the large FOV limit for the rain field model with a length scale, probably due to the neglect of higher order terms in the expansion. Short (1988) proposed a statistical model in examining the ESMR-5 (Electrically Scanning Microwave Radiometer on NIMBUS 5) data for rainfall retrieval. In his model, the whole ensemble of rainrates is transformed and hence can be compared with the large FOV limit in our case. He showed that rainrate inferred from the microwave brightness temperature underestimates the true rainrate by a factor that depends on the mean and variance of the rainrate distribution, a result that supports our formula.

   Much of the preceding is predicated on the assumption that a
horizontal length scale exists for rain fields. This assumption has
recently been questioned and the notion of fractals in rainrate
process has been introduced (Lovejoy and Mandelbrot, 1985; Lovejoy and
Schertzer, 1985). Analysis of raingage data, however, shows no
evidence of scaling for rainrate increments in time (Zawadzki 1987).
Kedem and Chiu (1987) argued that, due to the intermittent nature of
rain fields, they cannot be self-similar in a strict sense. It
appears that, in the satellite retrieval case, the question of
self-similarity down to essentially infinitesimal scales is avoided by
the fact that the microwave temperatures are results of integrating
vertically through the column of rain. In other words, a length scale
(the height of rain column) is naturally imposed in the retrieval
problem.

   We showed that the bias associated with nonuniformly filled FOVs
of microwave sensors is large but can be corrected. This bias depends
strongly on the rainrate variance over the FOVs. This term needs to
be estimated globally if the formulation is to be adopted in
algorithms of satellite rainfall retrieval. By invoking Taylor's
frozen field hypothesis, the equivalence of time and area averages has
been demonstrated in some cases (Zawadzki, 1975). Data from the
global network of raingages can be used to estimate the rainrate
variance term. Rain field models are also needed to enhance our
understanding of the physics of the bias associated with nonuniformly
filled FOVs of microwave sensors.

APPENDIX: SPECTRAL REPRESENTATION OF BIAS FORMULA

   We consider the Fourier representation of a one-dimensional rain
field

$$R(x) = (\sqrt{2\pi})^{-1} \int_{-\infty}^{\infty} \exp(ikx) \, R_k \, dk$$

where

$$R_k = (\sqrt{2\pi})^{-1} \int_{-\infty}^{\infty} \exp(-ikx) \, R(x) \, dx$$

and k is the wave number. Assuming a FOV of size a, the area averaged
rainrate is

$$[R] = 1/a \int_{-a/2}^{a/2} dx \; 1/(\sqrt{2\pi}) \int_{-\infty}^{\infty} \exp(ikx) \, R_k \, dk$$

$$= 1/(\sqrt{2\pi}) \int_{-\infty}^{\infty} R_k \, (\sin ka/2)/(ka/2) \, dk$$

Using the orthogonal relation for homogeneous statistics

$$\langle R_k \, R_{k'} \rangle = |R_k|^2 \delta(k-k')$$

where δ denotes the Dirac delta function and is zero except at $k=k'$,
it can be shown that

$$\langle [R]^2 \rangle = 1/2\pi \int_{-\infty}^{\infty} [\sin(ka/2)/(ka/2)]^2 \ |R_k|^2 \ dk$$

$$\langle [R^2] \rangle = 1/2\pi \int_{-\infty}^{\infty} |R_k|^2 \ dk$$

and hence

$$\langle [(R-[R])^2] \rangle = 1/2\pi \int_{-\infty}^{\infty} |R_K|^2 \ [1- ((\sin ka/2)/(ka/2))^2] \ dk \qquad (A1)$$

The term in brackets on the RHS acts as a high-pass filter on the spatial spectrum $|R_k|^2$. Hence if the spatial spectrum of rain is known, the bias can be readily evaluated.

REFERENCES

Arkell, R., and M. Hudlow, 1977: GATE International Meteorological Radar Atlas, US Dept. of Commerce, NOAA, US Government printing Office, Washington, D.C., 20402.

Arkin, P. A., 1979: The Relationship between Fractional Coverage of High Cloud and Rainfall Accumulation during GATE over the B-Scale Array, Mon. Weath. Rev., 107, 1382-1387.

Austin, P., and S. Geotis, 1980: Precipitation Measurements over the Oceans, Air Sea Interaction, F. Dobson, L. Hesse and R.Davis, Eds, Plenum Pub. Corp., 227 West 17th St., New York, NY, 10011.

Austin, P., and S. Geotis, 1978: Evaluation of the Quality of Precipitation Data from a Satellite-Borne Radiometer, NASA Report under Grant NSG-5024, Massachusetts Institute of Technology, Cambridge, MA 02139.

Atlas, D., and O. Thiele, 1981: Precipitation Measurements from Space, Workshop report, October, 1981, NASA, Goddard Space Flight Center, Greenbelt, MD 20771.

Barrett, E. C., and D. M. Martin, 1981: The Use of Satellite Data in Rainfall Monitoring, Academic Press, New York, NY, 340pp.

Chiu, L. S., 1988: Estimating Areal Rainfall from Rain Area, Tropical Precipitation Measurements, J. Theon and N. Fugono, (Eds.), Deepak Pub. Co.

Kedem, B., and L. Chiu, 1987: Are Rainrate Processes Self-Similar? Water Resour. Res., 23, 10,1816-1818.

Kummerow, C., 1987: Microwave Radiances from Horizontally Finite, Vertically Structured Precipitating Clouds, Ph. D. Dessertation, University of Minnesota.

Lovejoy, S., and B. Mandelbrot, 1985: Fractal Properties of Rain and a Fractal Model, Tellus, 37A,209-232.

Lovejoy, S., and D. Schertzer, 1985: Generalized Scale Invariance in the Atmosphere and Fractal Models of Rain, Water. Resour. Res., 21, 1233-1250.

Patterson, V. L., M. D. Hudlow, P. J. Pytlowany, F. P. Richardson and J. D. Hoff, 1979: GATE Radar Rainfall Processing System, NOAA Technical Memorandum, EDIS 26, NOAA, Washington, D.C.

Short, D. A., 1988: A Statistical-Physical Interpretation of ESMR-5 Brightness Temperatures over the GATE area, Tropical

   Precipitation Measurements, J. Theon and N. Fugono, (Eds.),
   Deepak Pub. Co.
Simpson, J., R. Adler, and G. North, 1988: A Proposed Tropical
   Rainfall Measuring Mission (TRMM) Satellite, Bull. Amer.
   Meteor. Soc., 69, 278-295.
Smith, E. A., and S. Q. Kidder, 1978: A Multispectral Satellite
   Approach to Rainfall Estimates. Unpublished manuscript.
   Colorado State University, Fort Collins, Colorado, 80523, 26pp.
   plus tables and figures.
Wilheit, T. T., A. T. C. Chang, M.S.V. Rao, E. B. Rodgers and
   J. S. Theon, 1977: A Satellite Technique for Quantitatively
   Mapping Rainfall Rates over the Oceans, J. Appl. Meteor., 16,
   551-560.
Wu, R., and J.A. Weinman, 1984: Microwave Radiances from
   Precipitating Clouds Containing Aspherical Ice, Combined Phase
   and Liquid Hydrometeors, J. Geophys. Res., 89, 7170-7178.
Zawadzki, I., 1987: Fractal Structure and Exponential Decorrelation
   in Rain, J. Geophys. Res., 92, D8, 9586-9590.
Zawadzki, I., 1975: On Radar-Raingage Comparison, J. Appl. Meteor.,
   14, 1430-1436.
Zwally, J. H., J. C. Comiso, C. L. Parkinson, W. J. Campbell,
   F. D. Carsey, and P. Gloerson, 1983: Antarctic Sea Ice
   1973-76: Satellite Passive Microwave Observations, NASA SP-459,
   NASA, Washington, D.C.

MICROWAVE REMOTE SENSING OF ATMOSPHERIC WATER VAPOR AND CLOUD LIQUID WATER OVER EQUATORIAL WESTERN PACIFIC OCEAN WITH A SHIPBORNE DUAL-WAVELENGTH RADIOMETER

Wei Chong, Lin Hai, Zou Shouxiang, Xuan Yuejian, Xin Miaoxin,
Wang Pucai and Lu Daren
Institute of Atmospheric Physics, Academia Sinica
Beijing, PRC

ABSTRACT

The microwave radiative transfer of a cloudy atmosphere at wavelengths of 1.35 cm and 0.85 cm was calculated with conventional radiosonde data at Guangzhou, China(23°N, 113.3°E) and the Yap Island, trust territory of the U.S.A. (9.6°N, 138°E). The regressive relations were established among microwave brightness temperature(T_b), microwave attenuation(τ), atmospheric column water vapor (Q) and liquid water content of clouds(L). A physically-statistical method is applied to retrieve the Q and L from T_b and τ based on these relations.

During December 1985–January 1986 and October–December 1986, the shipborne MW radiometers at wavelengths of 1.35 cm and 0.85 cm were jointed in the "Survey over Tropical Western Pacific," which is a part of the research project entitled "Ocean-Atmosphere Interaction over Western Pacific Area and Its Relation to Interannual Variation of Eastern Asia Climate," sponsored by Academia Sinica. Representative values of Q and L over vast equatorial Western Pacific area were obtained with the MW radiometer. Radiometer-measured Q were compared with

simultaneously launched radiosonde data with good agreements. Both
statistical results and case studies are given in this paper. The
ability of MW remote sensing in the research of ocean-atmosphere
interaction will be discussed.

1. INTRODUCTION

For the past few years the Institute of Atmospheric Physics
has been developing a dual-channel radiometer for remotely sensing
atmospheric water vapor and cloud liquid water content[Huang et
al., 1986a, 1986b, 1987; Wei et al., 1984]. The technique employs
two frequencies: one, near 22.235 GHz, is sensitive to water vapor,
while the second, at 35.3 GHz, is sensitive to liquid water
content. The ground-based dual-wavelength microwave radiometer has
been proved to be an effective approach to the real-time remote
sensing of atmospheric water vapor and cloud liquid water, which
are foundamental parameters to meteorology [Hogg et al., 1983;
Zhao et al., 1984; Huang, et al., 1987]. However, most of the
available ground-based measurements were taken in middle latitude
areas with continental climate. It is meaningful to expand the
application of this technology to the measurements of   the
atmospheric water vapor and cloud liquid water content over vast
ocean area where the meteorological observations are scarce.
Compared with satellite MW observations, the ground-based MW
observations have much less spatial coverage ability but have
higher area resolution and in principle, the ground-based MW
observations have better accuracy. Thus, it is expected that a
ship-borne MW radiometer will be an effecient technique to the
study of water budget and cloud radiation feedback and
supplementary to satellite remote sensing and conventional
meteorological observations.

Since 1985, Academia Sinica(Chinese Academy of Sciences) has
sponsored and organized a research project entitled "Ocean-Atmosphere Interaction over Western Pacific Ocean and Its Relation to

Interannual Variation of Eastern Asia Climate." The shipborne dual-wavelength MW radiometer was taken part in the campaign of ocean atmosphere interaction observation during Dec. 1985 to Jan. 1986 and October-December 1986. Based on the observation, useful data of atmospheric column water vapor and vertically integrated cloud liquid water content over tropical and subtropical ocean area were obtained. The retrieval method used in data processing was adopted from Huang et al. (1986b, 1987), with some modifications. In the original method, since one must assume the cloud temperature in order to estimate the liquid water content, it may result in a fairly large error in the liquid water content. Also an incorrect liquid water estimate affects the accuracy of the water vapor measurement. In this paper we have shown that the total microwave attenuations of the atmosphere are directly derived from the observed MW at two brightness temperatures. Because of avoiding the cloud temperature, the retrieved errors of the cloud liquid water content may decrease. Quantitative estimates of atmospheric water vapor retrieved from radiometric measurements have been shown to be more accurate when compared with simultaneously launched radiosonde data in the ship. Both statistical results and case studies are given in this paper. The preliminary observations by the shipborne dual-wavelength radiometer show that it is feasible to determine cloud liquid water content and atmospheric water vapor over this ocean area.

2. THE RETRIEVAL METHOD

The retrieval method used in present work is a physically-statistical scheme with the assumption of thin atmosphere approximation. The empirical relations between the atmospheric microwave brightness temperature(T_b) and the atmospheric total attenuation(τ), the background atmospheric attenuation(τ^a) and water vapor (Q), the cloud attenuation(τ^c) and cloud liquid water content(L) are developed by the microwave radiative transfer calculations of 1.35-cm and 0.85-cm wavelengths under the cloudy atmosphere model.

The all regressive coefficients are derived from historical radiosonde data, with the model clouds inserted into those heights where
the relative humidities exceed 85 per cent. We used 91 samples in
Guangzhou, China(23°N,113.3°E) and 245 samples in Yap Island, the
trust territory of U.S.A (9.6°N, 138°E) during winter months as the
typical values of the subtropical and tropical areas, respectively.

First, the total attenuations of atmosphere at these two wavelengths are obtained from the relation between τ and T_b measured by
the dual-wavelength radiometer system,

$$
\tau_1 = \begin{cases} (x + y\ T_{b1}) \cdot 10^{-4} & \text{when} \quad T_{b1} < 70 \text{ K} \\[2mm] (r + s\ T_{b1} + t\ T_{b1}^2) \cdot 10^{-4} & \text{when} \quad T_{b2} > 70 \text{ K} \end{cases}
$$

$$
\tau_2 = (n + v\ T_{b2} + w\ T_{b2}^2) \cdot 10^{-4}
$$

where subscripts 1 and 2 represent the wavelengths of 0.85 cm and
1.35 cm, respectively. Assuming that the contribution of atmospheric gases constituents to the total attenuation τ is a constant,
the cloud attenuation τ^c at the wavelength of 0.85 cm is obtained
from the relation between τ and τ^c.

The cloud attenuations we established at two wavelengths, 0.85
cm(τ_1^c) and 1.35 cm(τ_2^c), can be interrelated with the following
formulae [Zhou Xiuji et al., 1982]

$$
\tau_1^c / \tau_2^c = (\lambda_2 / \lambda_1)^{1.968}
$$

thus τ_2^c is obtained from above equation. Then, using the regressive relation between the background attenuation of cloudy days τ_2^a
and atmospheric column water vapor Q, the atmospheric column water
vapor Q is derived.

In order to improve retrieved accuracy of cloud liquid water

TABLE 1. THE REGRESSIVE RELATIONS AND THEIR COEFFICIENTS
         IN THE RETRIEVAL METHOD

Regressive Equations	Symbols	Regressive Coefficients & Deviation (σ)	
		Guangzhou Winter (91 cases)	Yap Winter (245 cases)
$T_1 = (x + y\ T_{b1}) \cdot 10^{-4}$ when $T_b < 70$ K	x	-243.80	-368.45
	y	42.96	44.14
	σ	0.00179	0.0302
$T_1 = (r + s\ T_b + t\ T_{b1}^2) \cdot 10^{-4}$ when $T_b > 70$ K	r	646.02	957.50
	s	15.17	8.098
	t	0.218	0.241
	σ	0.00815	0.0136
$T_2 = (u + v\ T_{b2} + w\ T_{b2}^2) \cdot 10^{-4}$	u	580.10	1001.8
	v	16.45	8.247
	w	0.216	0.234
	σ	0.0062	0.0067
$T_1 = A + B \cdot L$	A	0.098	0.127
	B	$2.50 \cdot 10^{-4}$	$2.57 \cdot 10^{-4}$
	σ	0.037	0.046
$Q = m + n\ T_2^a,\quad$ (g/cm^2)	m	0.0568	0.551
	n	13.29	12.77
	σ	0.078	0.144
$T_1^a = d + f\ Q$	d	0.0357	0.0376
	f	0.0180	0.0178
	σ	0.0019	0.0023

content in cloud, we should correct
the influence of atmospheric water
vapor on the attenuation of 0.85-cm
wavelength. In fact, although the
wavelengths near 0.85 cm are far
away from the 1.348-cm (22.235-GHz)
rotational lines of water vapor and
the 5-mm oxygen absorption band,
the influence of water vapor is
still significant [Olivero, 1984;
Hogg, 1979]. Shown in Figure 1 is
the background attenuation of
cloudy days τ^a as the function of
the total precipitable water
vapor. By substituting the
retrieved Q into the equation $\tau_1{}^a =$
$d + f Q$, a regenerate gas
attenuation $\tau_1{}^a$ is obtained instead
of the constant A in the equation
$\tau_1 = A + B \cdot L$. Thus, the correction
to L can be made from the above

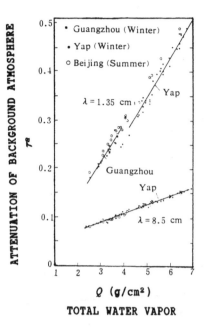

Q (g/cm²)

TOTAL WATER VAPOR

Figure 1. The relation between
total water vapor Q and the
attenuation of background at
wavelengths 8.5 mm and 1.35 cm.

equation. The regressive equations and their regressive coefficients in Guangzhou and Yap Island method are listed in Table 1.

3. FACILITIES AND OBSERVATIONS

        The instrument we used is a ship-borne dual-wavelength radio-
meter system consisting of two impulse noise-input null-balanced
Dicke-type microwave radiometers at wavelengths of 0.85 cm and 1.35
cm with same construction(see Figure 2). Specifications of the two
radiometers are shown in Table 2. The observation data are
automatically collected and processed by HP9825 computer. Sampling
time intervals are selected as 10 sec. and 1 min., respectively.
According to the in situ oceanic atmosphere environment during the
survey, the radiometers were calibrated either by a constant low

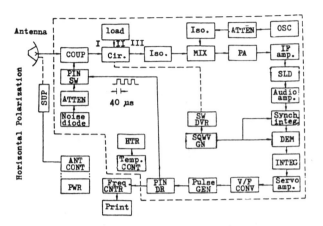

Figure 2. The block diagram of the radiometer.

temperature terminal load (liquid nitrogen of 77 K) in the laboratory or by comparing the instrument reading with the brightness
temperature calculated by the clear-air radiative transfer equation
in clear days with the radiosonde data in Guangzhou before the
survey.

For the past two years, we have carried out two cruises over
the Western Pacific Ocean. First cruise was from Dec. 1985 to Jan.

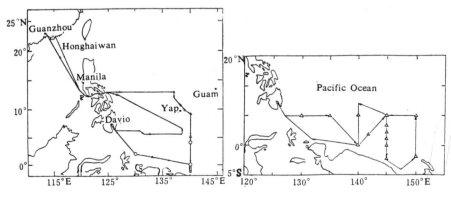

Figure 3. The ship track for
first cruise (Dec. 1985 -
Jan. 1986).

Figure 4. The ship track for
second cruise (Oct. - Dec.,
1986).

1986, and the second cruise was from Oct. to Dec. 1986. The ship
tracks for two cruises are shown in Figure 3 and Figure 4. The
atmospheric column water vapor and total liquid water content in a
variety of clouds in 6 stations for the first cruise and in 13
stations for the second cruise were obtained by the dual-wavelength
ship-borne microwave radiometer system. An infrared radiometer was
used for measuring zenith sky temperature. During cloudy days, it
can offer cloud base temperature. Total data of L and Q obtained
over tropical western Pacific Ocean have reached 5323 sets.

4. RESULTS

4.1 THE AMOUNT OF COLUMN WATER VAPOR AND ITS STATISTICAL CHARACTERISTICS

Because of strong convection and vast sea surface, there is
abundant water vapor over tropical western Pacific Ocean. Therefore, its measurement is significant for the research of tropical
Pacific water budget. Shown in Figure 5 is a comparison between
values of water vapor retrieved by the dual-wavelength radiometer
and those by real-time radiosonde released on board in cloudy and

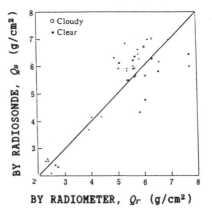

Figure 5. Comparison of total water vapor measured
by the dual-wavelength radiometer with that calculated from radiosonde data.

clear days over tropical Western Pacific during the two cruises. The correlation coefficient between the radiometry measurements and radiosonde is 0.876 and the error of root mean square is 0.734 g/cm^2. We have obtained the correlation coefficient of 0.99 for ground-based measurement [Wei et al., 1984]. However, considering the bad observation condition for both MW radiometer and radiosonde calibration over the ocean, in particular, the influence of ocean waves, we believe that the results are fairly satisfactory.

TABLE 2. THE RADIOMETER SYSTEM CHARACTERISTICS

Frequency		22.235 GHz(±200MHz)	35.3GHz
Antenna	Diameter	0.6 m	0.368 m
	Gain	40 dB	35 dB
	Main lobe	1.5°	1.5°
	Side-lobe level	−20 dB	−17 dB
Feed		Conical horn	
Total Noise Coefficient		9.4 dB	12 dB
Integral Time		3.3 sec.	1 sec.
Reference Load Temperature		48°C	48°C

The statistical characteristics of total water vapor in tropical atmosphere over Western Pacific Ocean are given in Figure 6, in which the left is relative frequency obtained from 3133 samples by using the dual-wavelength radiometer, and the right relative frequency obtained from 38 radiosonde data for comparison of each other. From this figure we can find that the measured values of Q of maximum occurences for both MW radiometer retrieved and

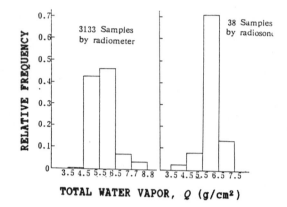

Figure 6. The relative frequency distribution of total water vapor over tropical Western Pacific Ocean.

radiosonde were located in the region of 5.5-6.5 g/cm² , which was almost twice the value for the middle latitude area where total water vapor was often about 3 g/cm². It should be noted that all data were obtained in winter months. Therefore, we can presume that total water vapor in the summer over Western Pacific must be more than above value.

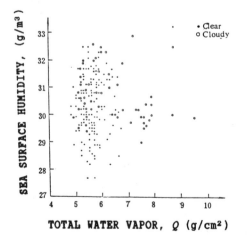

Figure 7. The relation of total water vapor with sea surface humidity.

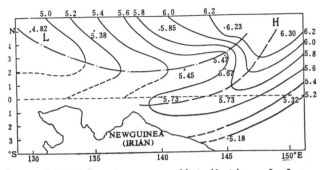

Figure 8. Total water vapor distribution of clear
days over tropical western Pacific Ocean.

As contrasted with that over land surface, the total water
vapor of atmosphere over ocean is not all dependent on sea surface
humidity. There is a complex relation between them. Figure 7
shows the distribution of total water vapor under clear and cloudy
days and its relation with sea surface humidity. The data were
obtained from 219 samples, of which the humidity at the sea level
have been simultaneously measured by a ventilated psychrometer on
the ship. The distribution range of total water vapor in clear days
is relatively narrow, as shown in this figure. The variation tendency consists with diurnal variation, shown in Figures 9 and 10.

4.2 SPATIAL DISTRIBUTIONS OF WATER VAPOR

Shown in Figure 8 is total water vapor distribution of clear
days selected from about one month observations over tropical
western Pacific Ocean, ranging from 130°E to 150°E and from 4°S to
5°N. From this figure, it can be found that the water vapor field
showed east-high and west-low feature and a wet tongue was
extending from the northeast to the equator. The dry air was from
the East Asia continent on the northwestern of this area. This
tendency is consistent with average multiannual distribution of
water vapor over the western Pacific Ocean.

4.3 DIURNAL VARIATION OF ATMOSPHERIC COLUMN WATER VAPOR
ON THE EQUATOR

A continuous observation of 24 hours over the Pacific Ocean near the equator (140°E) in Jan. 4 to Jan. 5, 1986 is shown in Figure 9. It can be seen that total column water vapor over the tropical ocean has much less diurnal variation except for transient increase during Cu cong's passing over the ship. This curve is consistent with the tendency of the variation of the water vapor pressure e at the sea level measured by a ventilated psychrometer.

Figure 10 is another example of diurnal variation for total column humidity over the equator of 150°E in November 17 to 18, 1986. Total water vapor reached to the maximum value at 3 p.m. and it appeared that the minimum value was reached at 2 a.m., with diurnal variation of 0.6 g/cm².

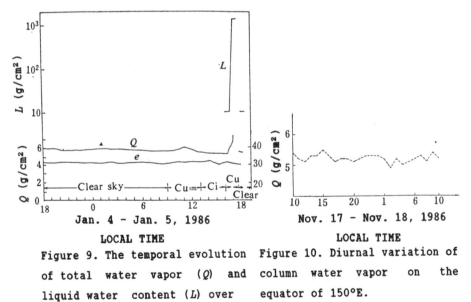

Figure 9. The temporal evolution of total water vapor (Q) and liquid water content (L) over the sea near the equator.

Figure 10. Diurnal variation of column water vapor on the equator of 150°E.

4.4 EVOLUTION FEATURES OF ATMOSPHERIC WATER VAPOR AND LIQUID WATER CONTENT UNDER TYPICAL WEATHER CONDITIONS

Figure 11-13 show the temporal evolutions of atmospheric total water vapor and cloud liquid water for different weather condition over the tropical Pacific Ocean. They were measured by the dual-wavelength radiometer about one hour before raining. Therefore we can find the evolution features between water vapor field and cloud liquid water when precipitating cloud bodies were just coming over the ship. In these figures, there is a common feature that, before

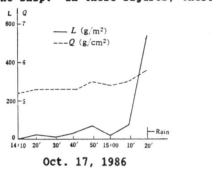

Oct. 17, 1986

LOCAL TIME

Figure 11. The temporal evolution of Typhoon cloud system (Sc op) (114°E, 23°N).

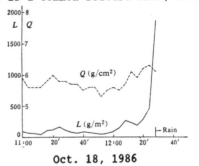

Oct. 18, 1986

LOCAL TIME

Figure 12. The temporal evolution of clouds.

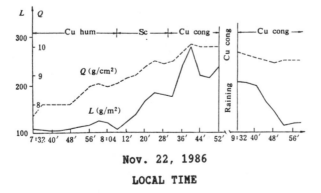

Nov. 22, 1986

LOCAL TIME

Figure 13. The temporal evolution of Cu cong (150°E, 5°N).

it is raining, both cloud liquid water and atmospheric water vapor increase rapidly, sometime accompanied with some irregular fluctuations. The phenomenon that cloud liquid water increases much more quickly in strong convective cloud has been observed many times, for instance, shown in Figures 11 and 12. However, in the condition of weak convective clouds, this tendency is not obvious; for instance, in Figures 13 and 14. More detailed structure, such as spatial distribution and long-term variation about them cannot be obtained from present surveys.

5. SUMMARY AND DISCUSSION

It has been proven that the microwave sensors can be used to estimate atmospheric column water vapor. Now the problem is how ground-based microwave radiometry is used in operational meteorological observation and some scientific research programs, such as nowcasting monitoring network, TOGA, and so on. Huang and Wei(1986a) proposed a feasible method for measuring layered water vapor contents, which would be a useful prediction factor of mesoscale weather nowcasting. The present results obtained over the equatorial western Pacific are representatives of that area. And the method described in above paragraphs can be applied to the observation of water vapor field and cloud LWC over both land and oceans. The study has also clearly shown that, under the condition of nonprecipiting clouds, the total atmospheric attenuations at two wavelengths can be obtained from the measurements of T_b directly, with at least the same accuracy as those by using the mean atmospheric radiant temperature. So it is feasible to determine of Q and L over the ocean.

As mentioned above, the total column water vapor and sea-surface humidity are required for the computation of water budget and heat fluxes, which are part of basic data in the research of air-sea interaction. The radiometer system can exist in a tropical upper air station on a small island or a scientific surveying ship

to continuously offer humidity data for the model and calibration
of satellite microwave sensor measurement of atmospheric column
water vapor.

The program on the survey over tropical western Pacific Ocean
is being carried out from 1985-1990, with more than five cruises.
The continuous measurements of total column water vapor data by the
dual-wavelength microwave radiometer, which has attended all
cruises, will help to reach an understanding of optimal remote
sensing techniques for estimating water budget and heat flux
components.

REFERENCES

Hogg, D.C. and F.O. Guiraud, 1979: Microwave Measurements of the
    Absolute Value of Absorption by Water Vapor in the Atmosphere,
    Nature, 279, 408-409.

Hogg, D.C., M.T. Decker, F.O. Guiraud, K.B. Earnshaw, D.A. Merritt,
    K.P. Moran, W.B. Sweezy, E.G. Strauch, E.R. Westwater and
    C.G. Little, 1983: An Automatic Profiler of the Temperature,
    Wind and Humidity in the Troposphere, J. Clim. & Appl.
    Meteor., 22, 807-831.

Huang Runheng and Wei Chong, 1986a: Experimental Investigation on
    Real-time Sensing of Layered Atmospheric Precipitable by a
    Ground-based Radiometer of 1.35 cm Wavelength, Advances in
    Atmopheric Sciences, 3, 86-93.

Huang Runheng and Wei Chong, 1986b: Remote Sensing of Atmospheric
    Water Vapor by Ground-based Microwave Radiometry, Presented
    at Beijing International Radiation Symposium, 26-30, August,
    1986, Beijing, China.

Huang Runheng and Zhou Shouxiang, 1987: Remote Sensing of Total
    Water Vapor and Liquid Content of Cloudy Atmosphere by Two-
    Wavelength Microwave Radiometry(in Chinese), Scientia Atmos-
    pherica Sinica, Vol 11, No.4, pp397-403.

Olivero, J.J., 1984: Microwave Radiometric Studies of Composition and Structure, Ground-based Techniques, Middle Atmosphere Program Handbook for MAP, Vol.13.

Wei Chong, Y. Xue, X. Zhu and S. Zou, 1984: Determination of Atmospheric Precipitable Water and Humidity Profiles by a Ground-based 1.35-cm Radio-meter, Advances in Atmospheric Sciences, 1, 119-127.

Zhao Bolin, Li Huixin and Han Jinyuan, 1984: Atmospheric Microwave Radiation and Remote Sensing of Water Vapor(in Chinese), Kexue Tongbao, No.4.

Zhou Xiuji, Lu Daren, Huang Runheng and Lin Hai., 1982: The Fundamentals of Atmospheric Microwave radiation and Microwave Remote Sensing, Science Press, Beijing, 178pp.

MICROWAVE RADIOMETER REMOTE SENSING
OF ATMOSPHERIC SOUNDING, OIL SLICK, AND SOIL MOISTURE

Zhao Bolin
Department of Geophysics, Peking University
Beijing, PRC

ABSTRACT

A set of microwave radiometers from 5-mm to 3-cm waveband of frequency (9.37 GHz, 35.3 GHz, 52.9 GHz, and 54.5 GHz) used in meteorology has been developed. The atmospheric remote sensing station with these microwave radiometers has been constructed. The efficiency of remote sensing of the atmosphere has been proven by field observation: (1) all-weather (except rain) remote sensing of atmospheric temperature, humidity, pressure, and liquid water content of cloud; (2) remote sensing of weather processes, for example, the monitoring of the change of air mass continuously, which may be used as a tool of newscasting; (3) detecting rainfall intensity by 3-cm radar combined with 3-cm radiometer. It could improve the precision in rain intensity measurement.

A laboratory for microwave remote sensing of earth substances has been constructed to improve the remote sensing environments. It includes dielectric constant measurement, microwave emissivity meter, reflectometer, and microwave remote sensing simulation experiment in the field, and it stressed study in remote sensing of oil slicks on water surfaces and soil moisture.

1. MICROWAVE RADIOMETER

The atmosphere is known to be a source of heat noise. The atmospheric absorption spectrum is shown in Fig. 1. By using atmospheric noise reception of 5-mm oxygen band, 1.35-cm water vapor band, 8-mm and 3-cm window bands, we could get the atmospheric temperature humidity profiles, total water vapor content, and liquid water content of clouds and rain. A set of microwave radiometers from 5-mm to 3-cm waveband of five frequencies (9.37 GHz, 22.235 GHz, 35.3 GHz, 52.9 GHz, and 54.5 GHz) used in meteorology has been developed. The atmospheric remote sensing station with these microwave radiometers has been constructed. Between 1970 and 1985, 12 types of microwave radiometers were developed. They are either Dicke's-type or compensative-type. The parameters of some radiometers are shown in Table 1.

MICROWAVE REMOTE SENSING
of the EARTH SYSTEM
Alain Chedin (Ed.)

127

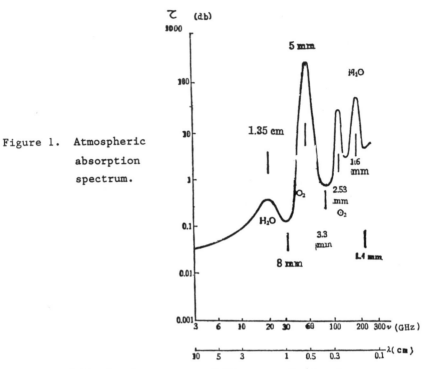

Figure 1. Atmospheric absorption spectrum.

Table 1. Parameters of Microwave Radiometers.

Wavelength	$\lambda = 5$ mm			$\lambda = 1.35$ cm	$\lambda = 8$ mm	$\lambda = 3$ cm
	Type A	Type B	Type C			
Frequency (GHz)	52.9	52.9	54.5	22.235	35.3	9.37
Radiometer type	Dicke's	compensative	compensative	compensative	compensative	Dicke's
Antenna	horn	Cassegrain parabolic	Cassegrain parabolic	Cassegrain parabolic	Cassegrain parabolic	J-feed parabolic
Antenna angle	$2\theta = 2.8°$	$2\theta = 1.5°$	$2\theta = 1.5°$	$2\theta = 1.5°$	$2\theta \sim 1.5°$	$2\theta = 1.5°$
Mixer	3db bridge	orthogonal field	orthogonal field	orthogonal field	orthogonal field	orthogonal field
Local oscillator	Shepard tube	solid state oscillator	solid state oscillator	solid state Oscillator	solid state oscillator	solid state oscillator
I. F. (MHz)	80	80	80	100	80	100
Bandwidth (MHz)	from 30 to 110	from 30 to 110	from 30 to 110	from 20 to 120	from 30 to 110	from 20 to 120
Noise figure (db)	16.5	10	11	7	9	7.2
Measurement parameter	temperature profile $T(z)$	temperature profile $T(z)$	temperature profile $T(z)$	humidity $q(z)$	cloud water contents W	rain water contents M
Height or distance (km)	<10	<10	<3	<10	<10	~10

2. MICROWAVE RADIOMETER REMOTE SENSING

2.1 TEMPERATURE PROFILE

The temperature profiles are derived by means of 5-mm radiometers with angle scanning observation. The examples of clear sky are shown in Figs. 2 and 3. The cloud effect may be neglected when the total liquid water content is small (i.e., smaller than 0.02 mm). The example of remote sensing of atmospheric temperature profiles in thin-cloud sky is shown in Fig. 4.

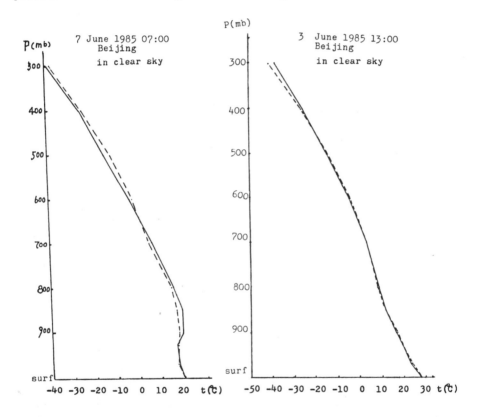

Figure 2. Example of microwave remote sensing of atmospheric temperature profile with inversion lapse by 5-mm (52.9 and 54.5 GHz) radiometers (Beijing).
—————— radiosonde
- - - - - radiometer

Figure 3. Example of microwave remote sensing of atmospheric temperature profile with superadiabatic lapse by 5-mm (52.9 and 54.5 GHz) radiometers (Beijing).
—————— radiosonde
- - - - - radiometer

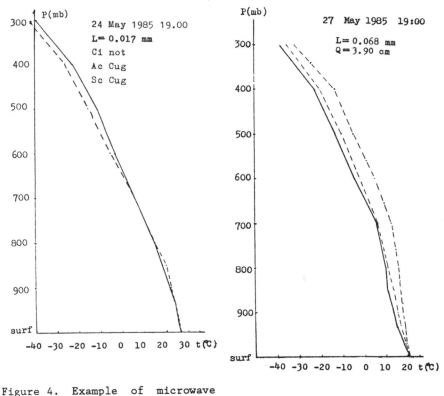

Figure 4.   Example of microwave remote sensing of atmospheric temperature profile of thin cloudy sky by 5-mm radiometer. (Beijing).

——————— radiosonde
- - - - - radiometer

Figure 5.   Example of microwave remote sensing of atmospheric temperature profile of cloudy sky with 5-mm, 8-mm, and 1.35 cm radiometers (Beijing).
L: liquid water content of cloud
Q: total water vapor content
——————— radiosonde
- - - - - radiometer (with cloud correction)
-- -- -- radiometer (without cloud correction)

The cloud effect needs correction with 8-mm and 1.35-cm radiometer zenith observation. Then the temperature profile may be derived by means of 5-mm radiometers with angle scanning observation. The example of remote sensing atmospheric temperature profile in cloudy sky is shown in Fig. 5.

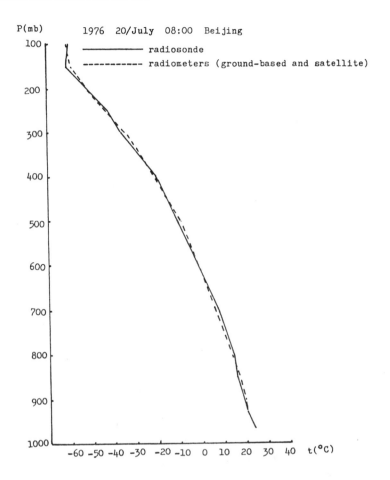

Figure 6.   Example of remote sensing of atmospheric temperature profile by ground-based radiometers combined with satellite radiometer observation ground-based radiometer: zenith brightness temperature of frequencies 52.9, 54.5, 22.235, and 35.3 GHz.
TIROS-N radiometer: Frequencies 50.35, 53.79, 55.02, and 58.01 GHz.

The temperature profiles can be derived by combining ground-based radiometer observation with satellite radiometer remote sensing. The example is shown in Fig. 6.

The variation of temperature profile may be derived from microwave remote sensing. The example is shown in Fig. 7.

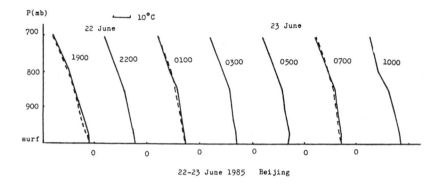

Figure 7.   The variation of temperature profile derived
            from microwave remote sensing, 22-23 June 1985
            (Beijing).   5-mm (52.9 GHz) observation with
            angle scanning, retrieval temperature by
            statistical and iterative method.
            _____ microwave radiometer
            - - - - - radiosonde

2.2 HUMIDITY

The atmospheric humidity profiles are derived by means of
1.35-cm radiometer with angle scanning observation in clear sky.
The example is shown in Fig. 8.

The total water vapor content Q and liquid water content
of cloud L may be derived by 1.35-cm and 8-mm radiometers with
zenith observation, in all weather conditions.   The results of
microwave remote sensing of total water vapor content, in
comparison with those obtained by radiosonde, are shown in Fig. 9.

3. MICROWAVE REMOTE SENSING OF WEATHER PROCESS

3.1 MONITORING THE CHANGE OF AIR MASS

There are three indices which can be measured: thickness
of 1000 mb-500 mb, total content of water vapor, and temperature
difference of 900 mb and 600 mb.

The results of microwave remote sensing of these
parameters in comparison with those obtained by radiosonde are
shown in Figs. 10, 11, and 12.

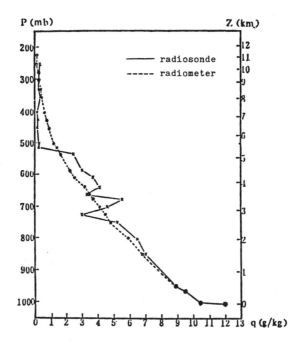

Figure 8.   Example of remote sensing of atmospheric
            humidity by 1.35-cm microwave radiometer,
            09:00, 28 June 1979 (Beijing) cloud (7) Cilfil
            Astra.

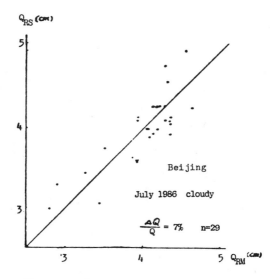

Figure 9.   The total water vapor content obtained by
            microwave radiometers Q_{RM} in comparison with
            the results obtained by radiosonde Q_{RS} (in
            cloudy sky).

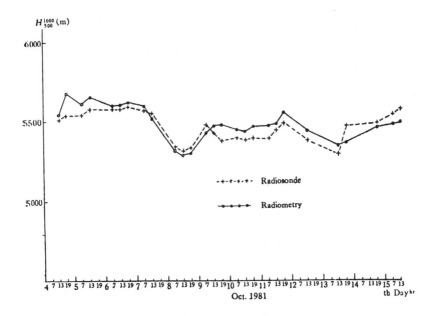

Figure 10. The variation of thickness between 1000-mb and 500-mb surfaces H_{500}^{1000} (Beijing).

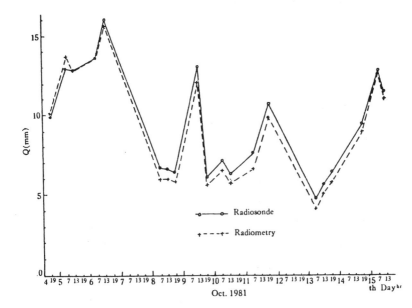

Figure 11. The variation of total water vapor content derived from microwave remote sensing (Beijing).

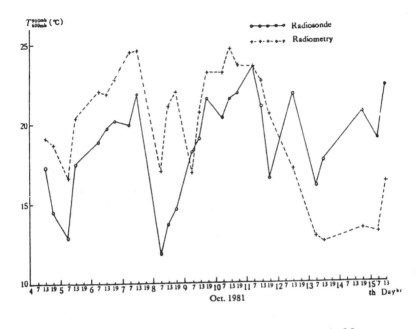

Figure 12. The variation of temperature difference between 900 mb and 600 mb T_{600}^{900} (Beijing).

3.2 MONITORING THE PRECIPITATION PROCESS OF CLOUD

Before rain begins, the total water vapor content and liquid water content increase rapidly. The radiometers can be used to monitor the change and it provides some information for nowcasting. The examples are shown in Fig. 13.

4. MEASUREMENT OF RAIN INTENSITY BY MICROWAVE RADIOMETER COMBINED WITH RADAR

The error of radar rain intensity measurement is mainly caused by the instability of radar constant and the uncertainty of raindrop spectrum. The rain emission, measured by radiometers is used to correct the radar constant instability and the influence of raindrop spectrum uncertainty. The developed 3-cm radiometer has been combined with Chinese-built 3-cm radar to detect the distribution of rain measurement. The results have been compared with those measured by rain gauge, the mean deviation between them is 25%. The rainfall variation of Miyun Reservoir Area of Beijing was measured.

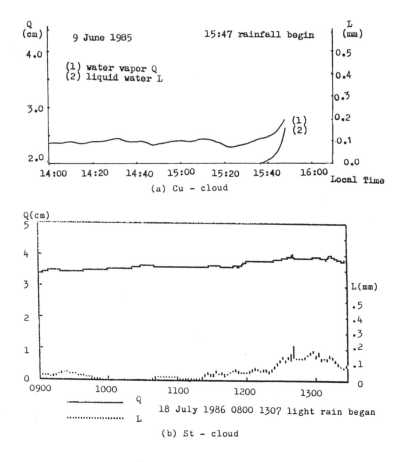

Figure 13.   The variation of total water vapor content Q
and liquid water content of cloud L (Beijing).

5. MICROWAVE REMOTE SENSING OF OIL SLICK
AND SOIL MOISTURE

5.1 A LABORATORY FOR THE REMOTE SENSING OF EARTH SUBSTANCE HAS BEEN CONSTRUCTED

In this laboratory a system of measuring the dielectric
constant of earth substances has been established, a reflectometer
and an emissivity meter have been developed, and some field
simulation experiments of remote sensing have been done.   The
structure of emissivity meter is shown in Fig. 15.

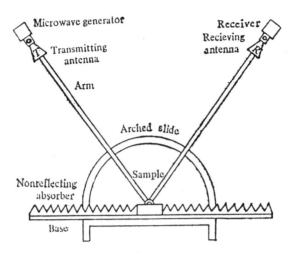

Figure 14. The structure of microwave reflectometer.

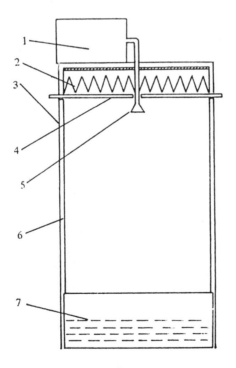

Figure 15. The structure of microwave emissivity meter.
1. microwave radiometer; 2. black body which
can be heated; 3. external part; 4. reflector;
5. horn antenna; 6. inner part, reflector; 7.
sample box

5.2 STUDY OF MICROWAVE REMOTE SENSING OF OIL SLICK ON WATER
    SURFACE

        The relationship between differently polarized
reflectivities with different angles and oil slick thickness
measured by reflectometer is shown in Fig. 16.

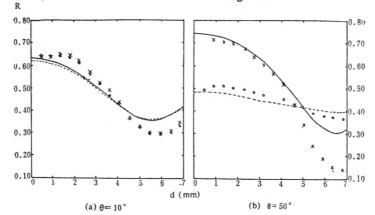

(a) θ= 10°   (b) θ= 50°

Figure 16. Microwave reflectivity against the thickness
            of oil slick (measured with reflectometer)
            (λ = 3 cm).

            ─────── Theoretical; x   experimental
                                  o

        The emissivity of oil slick on water surface by emissivity
meter measurement was carried out. The example is shown in Fig.
17.

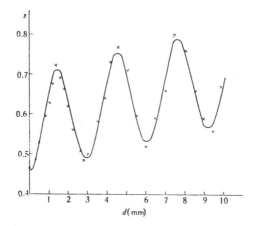

Figure 17. Microwave emissivity against the thickness of
            oil slick (measured with emissivity meter)
            (λ=8.5 mm, θ =0°).
            ─────── theoretical, + experimental

The differently polarized emissivity of oil slick on water surface by microwave radiometer in the field simulation experiment is shown in Fig. 18.

5.3 STUDY OF MICROWAVE REMOTE SENSING OF SOIL MOISTURE

The relationship between the microwave emissivity and water content of soil, and the effect of soil surface roughness on its microwave emissivity were measured. The results are shown in Fig. 19. The roughness parameter is h, $h=4\sigma^2 (2\pi/\lambda)^2$, where σ is the mean height of rough surface.

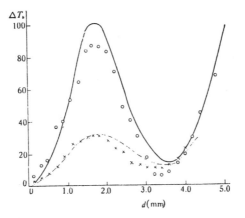

Figure 18.   Brightness temperature difference between oil slick and water against slick thickness (field experiment).
8.5 mm, θ =45°
Horizontal polarization
_____ theoretical, o experimental
Vertical polarization
- - - - -   theoretical, + experimental.

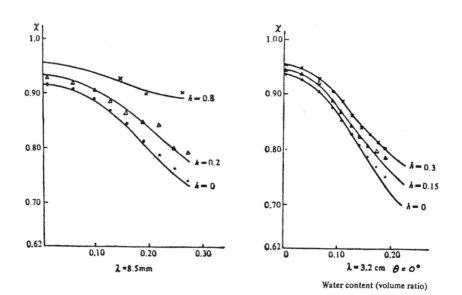

Figure 19. The microwave emissivity of soil with rough
            surface against soil moisture, measured with
            emissivity meter.
            _____ theoretical value
            experimental value: x   rough plain
                                 △   plain with ditch
                                 •   smooth surface

$$h = 4\sigma^2 (\tfrac{2\pi}{\lambda})^2, \quad \sigma \quad \text{mean height}$$

6. CONCLUSIONS

Microwave radiometry is a good method for: (1) remote sensing
of atmospheric sounding (i.e., temperature-humidity profiles, total
water vapor content, and liquid water content of cloud; (2) remote
sensing of weather processes (includes monitoring the change of
air mass and monitoring the precipitation of cloud); (3) remote
sensing of the thickness of oil slick on water surface and
monitoring the soil moisture.

SHIP-BASED PASSIVE MICROWAVE OBSERVATIONS IN THE WESTERN WEDDELL SEA DURING THE WINTER AND EARLY SPRING[1]

Thomas C. Grenfell
Department of Atmospheric Sciences AK-40
University of Washington
Seattle, Washington 98195, USA

ABSTRACT

Multifrequency passive microwave observations from the Weddell Sea during the austral winter of 1986 for 40 to 60 cm thick first-year ice shows four identifiable signature types, three of which have also been observed for northern hemisphere sea ice and one new type. This deficiency is thought to be a selection effect. The differences in these spectra are most pronounced at 90 GHz and should have only a small effect on existing satellite algorithms. The variations appear to be due to differences in brine volume at the snow/ice interface and structural variations in the snow layer. Cluster analysis for thin ice suggests that it may be possible to resolve young ice types and that, for a two-dimensional analysis, this would be best accomplished using observations at 18 GHz vertical polarization together with 90 GHz vertical polarization.

1. INTRODUCTION

The capability of making all weather year-round observations of sea ice in the polar regions has been available since 1973, with the advent of ESMR, and has been extended by SMMR. The imagery thus generated has shown considerable detail suggesting substantial variability in surface emissivity. Interpretation of the imagery has relied on observations made both from aircraft-mounted and surface-based instruments. Primarily for reasons of logistics, Arctic Sea ice has received considerable attention over the past 15 years, while no equivalent studies have been carried out in the Antarctic. Consequently, interpretation of microwave satellite imagery of Antarctic Sea ice has been based on the assumption that, at least on a small scale (centimeters to meters), it is the same as for first-year ice in the Arctic.

Recent measurements of the structure of Antarctic Sea ice, however, show that it is very different from the ice found in the Arctic Basin (Gow *et al.*, 1981; Comiso *et al.*, 1984). On the large scale, because the Antarctic ice is uncontained by surrounding land, its seasonal variation is much larger, and old ice exists only in a

---

[1] This project was made possible by a research grant from the National Aeronautics and Space Administration. We are also grateful for logistics support from the National Science Foundation.

MICROWAVE REMOTE SENSING
of the EARTH SYSTEM
Alain Chedin (Ed.)

141

few locations. Because of the much more severe weather, the produc-
tion of frazil ice in a wave field may be dominant, whereas in the
Arctic Basin, quiescent formation is much more prevalent. Consequently, even the crystal structure of southern sea ice can be
expected to differ significantly from that of its northern
counterpart.

In order to determine the characteristic emissivity spectra of
the various ice types and to see whether systematic differences exist
between similar ice types in either hemisphere, we took part in the
Western Weddell Sea Project (WWSP) under the sponsorship of the
Federal Republic of Germany and the US National Science Foundation.

2. EXPERIMENTAL PROGRAM AND OBSERVATIONS

From September through December 1986, during the austral winter
and early spring, we participated in the cruise designated WWSP V/3 on
the research icebreaker FS Polarstern. This was the first winter
transit by ship of the Antarctic ice pack up to and along the coast,
and provided a unique opportunity for comprehensive observations of
the winter sea ice. The project described here was the second part of
a six-month series of observations which spanned the early winter to
the onset of spring breakup. The area covered included four complete
transects of one of the world's widest seasonal sea ice zones.
Extensive sampling was carried out in the middle of the pack, and a
detailed examination of the near-shore polynya system was made extending from about 70° 30'S latitude, 8°W longitude to 77° 10'S, 34°W just
north of the Filchner Ice Shelf. Since initial results from the first
half of the winter have been presented by Comiso *et al.* (1988), most
of the information presented here will be from the period September
through December.

The general objective of the project was to provide ground-based
microwave data for comparison with SMMR and SSM/I observations, which
were directly related to the physical properties of the ice types
examined. We wished to determine the extent to which it would be
possible to identify and distinguish different ice types and/or
seasonal temporal changes and to compare the microwave signatures of
sea ice with those which have been observed in the north (Cavalieri *et
al.*, 1986; Grenfell, 1986; and Lohanick and Grenfell, 1986).

The observation program consisted of measuring brightness
temperatures of the ice at frequencies of 6.7, 10, 18.7, 37, and
90 GHz at both vertical and horizontal polarization. The radiometers
were mounted on the rail of the ship, 17 meters above the surface, in
a weathertight case which could be rotated in zenith/nadir angle to
observe the ice and the sky. When the ship was in motion, the nadir
angle was set to 50°, and traverse data were obtained alternating
between vertical and horizontal polarization in order to obtain a
large-scale record for comparison with satellite imagery. A more
detailed account of the instrument configuration, calibration, and
system characteristics is given by Comiso *et al.* (1988).

At more than 60 stations, occurring typically every two to three days, measurements of the physical properties of the ice were carried out. These consisted of direct measurements of temperatures at the snow surface and at the snow/ice interface, together with a series of cores through the ice for which the temperature profile was measured on site. The cores were then brought on board and the vertical profiles of salinity, density, and crystal structure were studied. Whenever snow was present, the density, mean grain size, and salinity profiles were also measured.

The ice conditions encountered ranged from a variety of young ice types, growing in both calm and wavy conditions, to white ice consisting of 40 to 60 cm thermal growth, covered with varying amounts of snow up to more than 20 cm in thickness. Persistent dynamical forcing from wind and ocean currents produced multiple overthrusting events in many places, resulting in total ice thicknesses of more than 2 m. These areas had rough surfaces due to protruding and broken ice blocks, and in these areas snowdrifts in excess of 50 cm were not uncommon. Even near the coast we encountered second or multiyear sea ice only in deep embayments in the ice shelf such as Drescher Inlet, and these were inaccessible to the ship. Although very thick ice, more than 7 m in some places, was encountered adjacent to the ice shelf, reports from ship's personnel from cruises during the previous summer indicated that this ice had all formed since that time.

3. RESULTS

Since the detailed reduction of the full data set is still in progress, most of the present discussion will be limited to selected cases of thick ice that comprised the larger ice floes studied at the ice stations. All of these cases have been selected from among ice types that were optically thick so that there was no direct influence of the underlying ocean on the emergent radiation and no direct effects due to differences in ice thickness.

Emissivities (e_i) at frequency i were derived from the brightness temperature (T_B) readings of the ice and the sky and the ice surface temperature (T_s) using Eq. (1).

$$e_i = (T_{B,i} - T_{sky,i})/(T_{s,i} - T_{sky,i}) \qquad (1)$$

For cases of snow-covered ice, $T_{s,i}$ was taken to be the snow/ice interface temperature. The emissivity spectra revealed the four distinct spectral types as shown in Figs. 1A through 1D.

For all cases the ice thickness ranged from 40 to 60 cm, and consisted predominantly of frazil ice with layers of congelation ice. In all cases the salinity was between 6 and 10 ppm, characteristic of first-year ice, and there was a layer of enhanced salinity at the snow/ice interface due to brine expulsion on cooling. Some of this brine was wicked up into the snow. Case A was very common throughout the experiment. It consisted of cold ice (<-10°C) covered with a 10-20+ cm layer of coarse grained windpacked snow. The grain diameter (d_{gr}) was greater than 1 mm, and the density was typically 0.4 gm/cm^3.

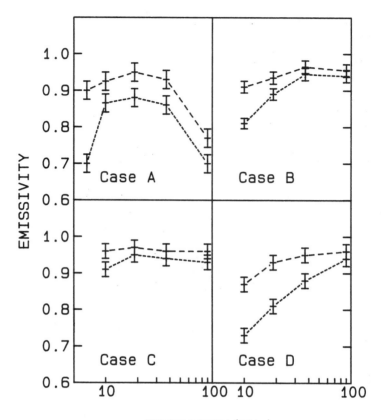

FREQUENCY (GHz)

Figure 1.   Emissivity spectra for four distinguishable cases of thick
            ice.   The long-dashed curves are for vertical polarization
            and the denser short-dashed curves are for horizontal
            polarization.

Case B was also frequently observed, again for cold ice, but the ice
was covered with only a thin-layer, low-density snow (0.1-0.15 gm/cm^3)
with a typical grain diameter of 0.3 mm. Case C was observed later in
the season when the ice was warmer but had a thick layer of fine-
grained snow (d_{gr} typically 0.3 mm).   Since the snow/ice interface
temperature, $T_{s/i}$, was still -6°C, little melting was taking place,
but the mean brine volume in the saline layer of the snow was considerably larger than for cases A and B, as shown in Fig. 2. Assuming
the snow was in eutectic equilibrium with a typical salinity of
35 ppm and using the formulae of Frankenstein and Garner (1967), we
find that at the interface v_{br} is approximately 32% for C as opposed
to about 20% for cases A and B.   Case D was observed in the early
spring as the ice was approaching its melting point, and was covered
by a moist snow layer with coarse metamorphosed grains.   Schematic
diagrams of the ice structure for these four cases are shown in
Figs. 2A through 2D.

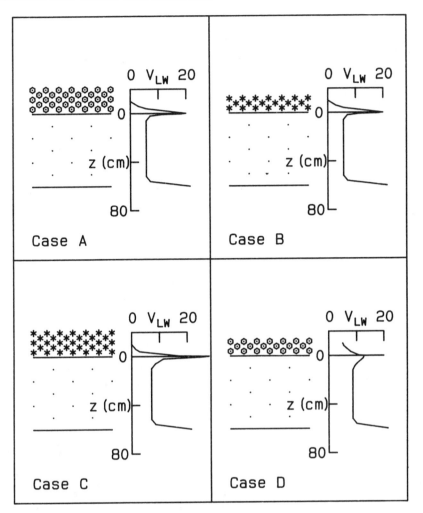

Figure 2. Schematic display of the ice and snow characteristics
corresponding to the spectra shown in Fig. 1. The dots
indicate the sea ice. The hexagons represent high-density
(about 0.4 gm/cm^3), coarse-grained snow (d_{gr} about 1 mm),
and the asterisks represent low-density (about 0.1 gm/cm^3)
fine-grained snow (d_{gr} about 0.1 mm). V_{LW} denotes the
volume fraction in percent of brine and/or liquid water
through the ice and snow. The curves are typical values
ignoring diurnal variations due to temperature and incident
radiation. The sharp rise at the bottom of the ice
represents the transition from ice to ocean water at 100.

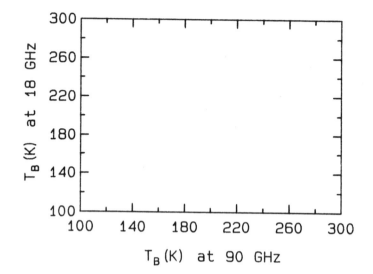

Figure 3.   Cluster diagram of brightness temperature at 18 GHz V-Pol
            versus brightness temperature at 90 GHz V-Pol for a single
            day's ship traverse. The horizontally extended uppermost
            group is from thick (40-60 cm) ice with varying amounts and
            densities of snow structures and snow/ice interface condi-
            tions; the very small group at (215,145) is open water; and
            the complex clustering in between is due to a variety of
            thin and young ice types.

        A case from the traverses of special interest is shown in the
cluster plot of Fig. 3.   This was taken when the ship was passing
through relatively large areas of thin ice. The combination of 18 vs.
90 GHz has been selected because it optimizes the resolution of ice
types. Most of the points fall in the upper cluster and correspond to
thick ice with varying degrees of snow cover. Also clearly resolved
are four or five smaller but distinct clusters corresponding to large
areas of new and young ice with different ice and snow thicknesses.
Ice thickness differences are due both to thermal growth and to overthrusting. Preliminary examination suggests that the separation in
the abcissa is due to differences in snow conditions while the separation in the ordinate is due to ice thickness.   Without the 90 GHz
data, these clusters would merge into a single secondary cluster.
This appears to correspond to the results of aircraft observations
over the Bering Sea (Cavalieri et al., 1968).

4.   CONCLUSIONS

        The spectra for cases B, C, and D are very similar to results
obtained in the north. Cases B and C, the former showing a positive
spectral gradient and the latter being spectrally flat, can be
compared directly with results from the Bering Sea (Grenfell, 1986),
where the snow was also very light and both the salinity and thickness

of the ice were about the same as for the present cases. The spectrum
of case D is typical of melting snow-covered ice and similar values
have been reported extensively from spring and early summer observations (Grenfell and Lohanick, 1985; Onstott et al., 1988).

The spectrum of case A has not been reported before. It is
unusual because, instead of increasing continually with frequency, the
emissivity typically shows a maximum at 18 or 37 GHz and a distinct
decrease at 90 GHz. This behavior did not appear to correlate with
rafting, but rather corresponds to a thick dense snow cover that was
not encountered during our Bering Sea observations. It is clear from
the upper cluster in Fig. 3 that the decrease at 90 GHz can cover
quite a wide range and is accompanied by almost no variation at
18 GHz. This is probably due to variations in the degree of scattering in the snow layer, which would, in general, be expected to cover
a continuous range. This appears to be the case evident in Fig. 3.
From the present experiment, it has become clear that this spectrum is
quite common. It is probably fortuitous that we have not observed it
previously in the north. This will be checked in future experiments.
The differences seen here for the thick first-year ice are most
evident at 90 GHz. As a result, they should not strongly affect
currently operational satellite algorithms for ice concentration and
ice type distribution, since these use values at 18 and 37 GHz.

An important result here is the potential separation of young
ice types using a combination of the 18 and 90 GHz channels, and the
same potential is true for the pair of 18 and 85 GHz now available on
SSM/I. The variation at 90 GHz does not appear to be correlated with
that at 18 GHz, thus 90 GHz seems to provide independent information
on thin-ice type. At this stage in the data analysis, however, this
phenomenon has been observed for only one day's observations. If this
pattern holds in general, it suggests that 90· GHz and also the high
frequency channel on SSM/I may be useful for determining the thickness
distribution of young ice types. A more definitive prediction will be
available when the entire data set has been processed.

ACKNOWLEDGMENTS

Our sincere thanks to the Alfred Wegener Institut and to officers
and personnel of FS Polarstern for continued logistics support. We
are especially indebted to Mr. Hajo Eiken and Dr. Michael Spindler,
who organized and carried out the ice structure analysis, and to Dr.
Steve Ackley for badly needed financial assistance.

REFERENCES

Cavalieri, D.J., P. Gloersen, and T.T. Wilheit, 1986: Aircraft and
satellite passive microwave observations of the Bering Sea ice cover
during MIZEX West, IEEE Trans. Geosci. Rem. Sens., GE-24, 368-377.

Comiso, J.C., S.F. Ackley, and A.R. Gordon, 1984: Antarctic sea ice
microwave signatures and the correlation with in situ observations, J.
Geophys. Res., 89, 662-672.

Comiso, J.C., T.C. Grenfell, D.L. Bell, M.A. Lange, and S.F. Ackley, 1988: Passive microwave in situ observations of winter Weddell Sea ice, J. Geophys. Res., in press.

Frankenstein, G., and R. Garner, 1967: Equations for determining the brine volume of sea ice from -0.5°C, J. Glaciol., 6, 943-944.

Gow, A.J., S. F. Ackley, W.F. Weeks, and J.W. Govoni, 1981: Physical and structural characteristics of antarctic sea ice, Ann. Glaciol., 3, 113-117.

Grenfell, T.C., 1986: Surface-based brightness temperatures of sea ice in the Bering and Greenland Seas, IEEE Trans. Geosci. Rem. Sens., GE-24, 378-382.

Grenfell, T.C., and A.W. Lohanick, 1985: Temporal variations of the microwave signatures of sea ice during the late spring and early summer near Mould Bay NWT, J. Geophys. Res., 94, 5133-5144.

Lohanick, A.W., and T.C. Grenfell, 1986: Variations in brightness temperature over cold first-year sea ice near Tuktoyaktuk, Northwest Territories, J. Geophys. Res., 94, 5133-5144.

Onstott, R., T.C. Grenfell, C. Matzler, C.A. Luther, and E.A. Svendsen, 1988: Evolution of microwave sea ice signatures during early and mid summer in the marginal ice zone, J. Geophys. Res., in press.

DUAL DOPPLER RADAR ANALYSIS OF THE MARINE MIXED LAYER DURING A COLD AIR OUTBREAK OVER A STRONG SEA SURFACE TEMPERATURE GRADIENT

Rob E. Marshall and Sethu Raman
North Carolina State University
Raleigh, NC 27695–8208, USA

ABSTRACT

An investigation of the structure of the marine boundary layer using metallic half wavelength chaff and dual Doppler radars is described. The measurements were made in cold air flow over a warm and complex sea surface temperature structure. Radar derived mixed layer height contours, vertical wind contours, divergence fields and eddy wind fields are displayed.

1. INTRODUCTION

The use of half wavelength metallic chaff as a radio frequency energy scatterer was an early facet of electronic counter measures. The relationship between the chaff density and radar cross section is given by Schlesinger (1961). Chaff is typically 25-micron diameter aluminum coated mylar thread cut to near one half the wavelength of the observing radar. The terminal velocity of chaff at centimeter radar wavelengths is 20 to 30 cm/sec. Coherent radar measurements of radial velocity in chaff are not contaminated by the relatively large terminal velocities associated with hydrometeor targets. Recently there has been a resurgence in the use of chaff by radar meteorologists studying the planetary boundary layer (PBL) in clear air. A thorough review of atmospheric boundary layer radar experiments is given by Kropfli (1983). This paper describes single and multiple Doppler radar PBL experiments which employed intrinsic clear air scatterers (turbulent or inhomogeneous media) as well as half wavelength chaff. Dual doppler radar analysis calculates the three-dimensional cartesian wind field from simultaneous measurements of radial velocity in the same boundary layer volume by two coherent radars. Theses calculations are closed by a vertical integration of the mass continuity equation. To date there have been two dual Doppler radar chaff experiments which observed the three-dimensional mesoscale velocity structure of the marine atmospheric boundary layer (MABL). The results of a dual Doppler radar chaff experiment over the Santa Barbara Channel in California are given by Kropfli and Wilczak (1986). This paper describes the mesoscale horizontal velocity field in a stable MABL as measured by two 3-cm wavelength coherent radars on a 42-km baseline.

A mesoscale MABL experiment was conducted off the Outer Banks of North Carolina on March 2, 1986, during the Genesis of Atlantic Lows Experiment (GALE). The field phase of GALE was from 15 January to 15 March 1986. The experimental area stretched from the Gulf Stream across coastal Georgia, South Carolina, North Carolina, and Virginia to the Appalachians. The objectives of GALE were to study the mesoscale and air-sea interaction processes in East Coast winter storms. Improved temporal and spatial measurements were made throughout the GALE measurement area with numerous soundings, surface measurements, ships, aircraft, radars, and satellite imagery. Details of GALE are given by Dirks et. al. (1988). One of the specific tasks of GALE was to study the three-dimensional mesoscale structure of the atmospheric

MICROWAVE REMOTE SENSING
of the EARTH SYSTEM
Alain Chedin (Ed.)

boundary layer and the associated energy and momentum fluxes over an area ranging from the Gulf Stream to the Appalachian Mountains approximately 500 kilometers from the coast. The planetary boundary layer subprogram of GALE employed specialized offshore observing platforms, research ships, research aircraft, Doppler radars, meteorological towers, mobile surface observations teams, soundings, and portable automated surface measurement networks. Raman and Riordan (1988) describe the boundary layer sub program of GALE.

In order to study the mesoscale structure of the MABL around the meteorologically active Cape Hatteras area, a dual Doppler radar experiment was designed. Half-wavelength chaff was dispersed in the clear MABL between 1320 and 1600 GMT on March 2, 1986. A schematic representation of the radar locations and dual Doppler coverage areas are shown in Fig. 1. The chaff was released along the line indicated by Fig. 1 over the Pamlico Sound. The chaff advected in northwest flow from the Pamlico Sound, over the barrier islands of Ocracoke and Hatteras, to the Atlantic shelf water and on to the warmer sea surface temperatures (SST) of the Gulf Stream.

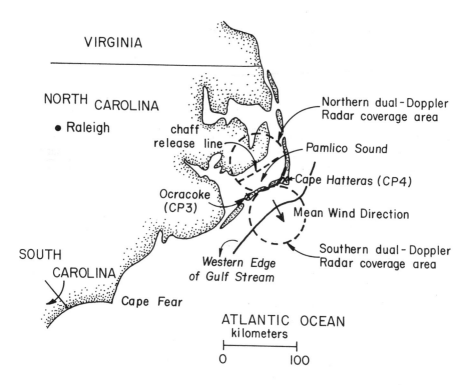

Figure 1.   GALE dual Doppler research area. The chaff plane flew 500 m ASL. The mean wind direction is for the March 2, 1986, experiment described in this paper.

The National Center for Atmospheric Research (NCAR) coherent pulsed radar number 4 (CP4) was located on Hatteras Island (75.53 deg W, 35.23 deg N). Another Doppler radar, CP3, was located on Ocracoke Island (75.97 deg W, 35.10 deg N). Radial velocity and reflectivity were measured in the advecting chaff cloud by the two coordinated 5.5-cm wavelength radars. The performance characteristics of these two radars is given in the GALE Experiment Design Document (1985). The baseline separation between CP3 and CP4 was 42 km. The northern dual Doppler coverage area was used to make measurements over the Pamlico Sound. The southern dual Doppler coverage are was used to make measurements over the Atlantic shelf water and Gulf Stream. A limitation of the dual Doppler techniques is that the boundary layer volume immediately above the radar baseline is unanalyzable.

As Fig. 1 indicates, the two radars are in a region characterized by the marine boundary layer. The Outer Banks of North Carolina are narrow strips of land 0.5 to 3.0 km in width. These strips of mostly sand are disected by several inlets which have come and gone over the centuries with hurricanes and winter cyclones. Hatteras Inlet bisects the radar baseline. The Outer Banks separate the wide shallow waters of the Pamlico and Albemarle Sounds from the Atlantic shelf waters and Gulf Stream. (The 16th century navigator Giovanni da Verrazzano believed upon viewing the Outer Banks that they separated the Atlantic Ocean from the Pacific!) The schematic representation of the western edge of the Gulf Stream in Fig. 1 indicates a seasonal average of the 20 deg c sea surface temperature (SST). Detailed SST contours derived from NOAA-7 imagery for March 2, 1986, will be presented in a later section.

The chaff plane flew at 500 meters above sea level (ASL) along the line indicated in Fig. 1 as the chaff was exhausted from a 4-inch diameter tube. This chaff line mixed down to the surface and up to the top of the boundary layer in approximately 40 minutes. Radar observations were made in this chaff cloud as it advected in northwest flow. This paper describes the mesoscale structure of the MABL using radar measurements made in the chaff cloud when it was near the western edge of the Gulf Stream at 1456 GMT on March 2, 1986.

2. SURFACE OBSERVATIONS

The surface weather map at 1200 GMT for March 2, 1986, is shown in Fig. 2. The boundary layer experiment discussed in this paper represents the concluding measurements of GALE Intensive Observation Period number 11 (IOP #11). This IOP produced extensive measurements over the Atlantic off the coast of North Carolina during a case of moderate offshore cyclogenesis. The synoptic scale surface analysis in Fig. 2 indicates that the low pressure system located off the coast of North Carolina the previous day (1 March 1986) had moved to 40.5 deg N and 64.0 deg W. A quiescent surface trough was located parallel to the radar baseline. A high pressure system was analyzed west of Florida at 26.0 deg N and 87.0 deg W. Both radar sites and the chaff pilot reported clear skies. The temperature and dewpoint reported by the National Weather Service (NWS), Hatteras station, were 2 and -6 deg C respectively. The winds were 8m/sec from a northwesterly direction (330 deg).

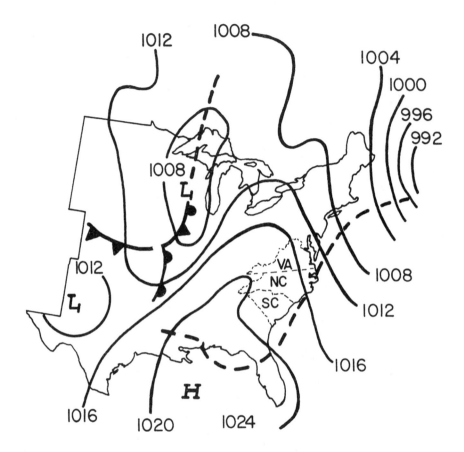

Figure 2.   Synoptic weather pattern at 1200 GMT on March 2, 1986. The GALE dual
             Doppler research area is at the coast of North Carolina. The sky over the
             research area was clear during the dual Doppler experiment period. Pressure
             is in millibars.

     Surface mesoscale observations for 1500 GMT were provided by the NCAR
Portable Automated Mesoscale Network (PAM). The PAM Network is described by
Dirks et al. (1988). The locations of the PAM stations near the radar measurement area
are shown in Fig. 3. Temperature and dew point in degrees centigrade are indicated
along with the wind vector at each PAM station. Notice the relatively strong winds near
Cape Hatteras. This is due mainly to the long fetch over the smooth Pamlico Sound
during this type of flow. In the last three years, Hatteras natives have named this area
of the Pamlico Sound "Canadian Hole" because of the spring influx of wind-surfing
Canadians and New Englanders. These data indicate that there was considerable
moistening of the boundary layer as it advected across the Pamlico Sound with no
noticeable change in the air temperature. Fig. 4 represents the 35 km gridded eddy wind
field produced from the PAM data. This plot is generated by subtracting the 15-minute
mean of all the PAM station winds from each PAM station wind vector. A northeast
eddy wind component over the northeast end of the Pamlico Sound and a more northerly
component over the southeast Pamlico Sound can be seen in Fig. 4. As will be shown

later, this wind field caused the initial chaff lie to bend near the middle as it advected over Pamlico Sound. The maximum air to sea surface temperature difference over the Pamlico Sound was 5 deg C with the water warmer than the air.

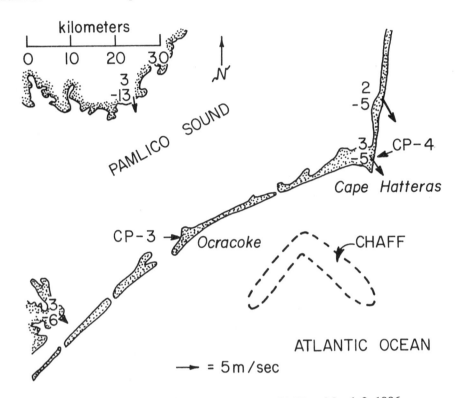

Figure 3.   Surface mesoscale observations at 1500 GMT on March 2, 1986. Temperature and dew points are in degrees centigrade. The chaff cloud location at 1456 GMT is shown. Radar derived wind fields in this chaff cloud at 1456 GMT are described in this paper.

A dual Doppler analysis of the chaff cloud over the Pamlico Sound indicated a highly turbulent and well mixed MABL up to 2.1 km. On the Atlantic Ocean side of the narrow Outer Banks, the shelf water surface temperatures drop to as low as 5 deg C. The air-sea temperature difference over the Atlantic shelf was reduced from 5 to 2 deg C. The resulting MABL was therefore more shallow with the mixed layer heights down to 800 meters. As the chaff cloud advected towards the Gulf Stream, the SST field became more complex with horizontal temperature gradients as high as 2 deg C per kilometer.

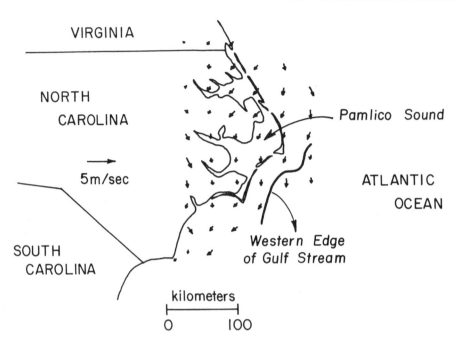

Figure 4. Eddy wind field at 1500 GMT on March 2, 1986. Grid points are 35
kilometers apart. Data were acquired from the NCAR PAM network.

3. MARINE BOUNDARY LAYER STRUCTURE OVER THE GULF STREAM

The MABL investigated during this experiment was expected to be influenced by
baroclinic effects introduced by the large temperature gradients associated with the Gulf
Stream. Arya and Wyngaard (1975) predict that the actual wind shear in the bulk of the
mixed layer will be smaller than the magnitude of the imposed geostrophic wind shear.
The magnitude of the geostrophic wind shear is approximately proportional to the
horizontal temperature gradient. Calculating the geostrophic shear involves evaluating
the thermal wind equations. Calculations of geostrophic shear over the Atlantic shelf to
Gulf Stream mesoscale area using the surface SST data produce very large values of
geostrophic shear up to 50 m/sec/km. These surface temperature gradients generally
diminish rapidly with height in the well mixed MABL. Computations of the geostrophic
shear using horizontal surface air temperature gradients would produce more reasonable
values of baroclinicity.

The increasing thermal instability associated with cold air advecting over a positive
SST gradient was expected to enhance the formation of the helical roll vortex. This
well-known coherent boundary layer structure was reviewed by LeMone (1973). The
neutral boundary layer model given by Brown (1970) indicates a stable to helical roll
circulation transition for mean winds greater than 7 m/sec. Mean wind speeds in excess
of 7 m/sec were estimated over the Pamlico Sound and Gulf Stream during post-frontal
northwest flow.

Sea Surface Temperature contours as observed by the polar orbiter NOAA-7 are
shown in Fig. 5. The horizontal resolution is 4 km. Notice the strong SST gradient
southeast of the radar baseline. The 10 deg C and 20 deg C SST contours are located at

strong SST gradients that loosely define the cross-current dimensions of the Gulf Stream. Preliminary data from airborne microwave radiometer measurements of SST's made along the mean wind direction indicate the 10-and 20-deg C contours are approximately 13 and 30 km offshore respectively. Small-scale eddies in the Gulf Stream between 13 and 30 km offshore are indicated by these preliminary airborne radiometer data as well as by in situ measurements made by the research vessel Cape Hatteras (Raman and Riordan, 1988). The general SST structure between the 10-deg C and 20-deg C SST contours along the mean wind direction is thus a step function with smaller scale, sometimes quasi-periodic, variations riding on the 17 km step function. The SST gradient at the 10 to 20 deg C step is as high as 10 deg C per kilometer in some locations. The advecting chaff cloud encountered a complex SST field which produced baroclinicity in the surface layer of the MABL. The Gulf Stream eddies introduce the possibility of the surface layer thermal wind equations switching polarity as the chaff cloud advected across the Gulf Stream.

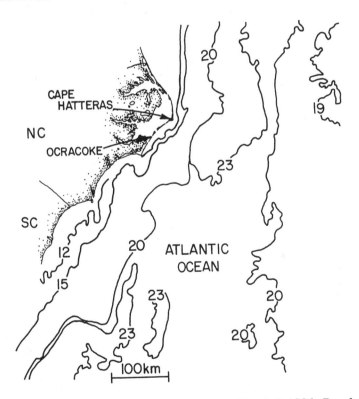

Figure 5.   Sea surface temperature contours on March 2, 1986. Resolution of the NOAA-7 data is 4 km.

Topography of the height of the marine boundary layer as it appears over the northwest edge of the Gulf Stream is shown in Fig. 6. Most of the chaff cloud is between the 10-deg C and 20-deg C SST contours. The rounded apex of the chaff cloud was between 24 and 32 km east of Ocracoke. The leading edge of the chaff cloud is characterized by areas of steeper topography in the mesoscale mixed layer height field. Mesa-like structures in the mixed layer height field in Fig. 6 are noticeable along the leading edge. The mesa located between 18 and 22 km east of Ocracoke and between 30 and 32 km have a wavelength of 2 km and a peak to peak amplitude of 1 km. The radar derived convergence at 100 meters ASL associated with these mesa was 0.002/sec. This is the order of magnitude for convergence one would expect in strong convection. Typical values of wind field divergence are given in Huschke (1980). The lagging edge of the chaff cloud has a more sloping mixed layer height field. The along mean wind wavelength of the mixed layer height variations is on the order of 4 km. Observations of the boundary layer height profile were also made over the measurement area by the NASA Electra research aircraft. The downward-looking lidar data indicate the same 4 km along mean wind wavelength for the mixed-layer height field (Boers, 1987). Notice the 700 meter trough in the mixed-layer height field between 21 and 23 km east of Ocracoke as described in Fig. 6. The measured radar reflectivity is proportional to the concentration of chaff (chaff fibers/volume). The chaff concentration is generally highest along the horizontal axis of the chaff cloud. This gives the impression that the horizontal mixing occurs in both directions along the mean wind axis. This axial chaff concentration core is discontinuous across the 700 meter ASL trough. A 4 to 1 chaff concentration variation is observed throughout the chaff cloud axis vertical plane across this trough in the mixed-layer height field. This implies mixing isolation in the axial direction between the two parts of the chaff cloud. The northwest and southeast ends of this trough are associated with radar derived convergence at 100 m ASL of .002/sec. The radar-derived eddy flow field shown in Fig. 9 indicates strongest southeast inflow for the chaff cloud to be associated with this trough. Southeast eddy winds up to 8 m/sec are found at the southeast side of the trough from 100 to 400 m ASL.

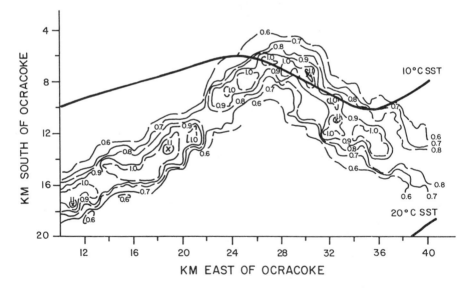

Figure 6.   Mixed layer height contours at 1456 GMT on March 2, 1986. Recall the location of this chaff cloud at 1456 GMT in Figure 3. The axes of this figure are measured relative to CP3 located at Ocracoke. The 10 and 20 degree centigrade sea surface temperature contours are indicated. Height contours are in kilometers. For example, the mixed layer height 13 kilometers south of Ocracoke and 19 kilometers east of Ocracoke is 1.1 kilometers.

The approximately constant 600-meter height in the mixed-layer field along the lagging edge of the chaff cloud in Fig. 6 appears to have been produced by strong outflow from 22 km to 27 km east of Ocracoke. The horizontal eddy wind analysis to be discussed below shows strong outflow along this trailing edge up to 500 m ASL. Radar-derived vertical wind profiles indicate this area of the chaff cloud is dominated by downdrafts with core velocities up to 0.7 m/sec. These downdrafts are on the order of 2 km across in the horizontal dimension. The updraft associated with this trailing edge is only 1 km across and has vertical core velocities near 0.5 m/sec.

Fig. 7 is the radar-derived vertical wind contours in the horizontal plane at 300 m ASL for the 1456 GMT scan. At this height the large area of downdraft associated with the trailing edge is easily seen. The 0.2 m/sec updraft along this trailing edge probably causes a one kilometer peak in the mixed layer height field. The updraft areas along the leading edge from 18 to 22 km east of Ocracoke are associated with the leading edge mesa structures in the mixed-layer height field shown in Fig. 6. The 0.6 m/sec updraft centered 7.0 km south and 31.0 km east of Ocracoke contributes to the 1.1 km peak height in the mixed layer located at 7.5 km south of Ocracoke and 30.0 km east of Ocracoke. The downdraft immediately to the southeast results in an area of mixed-layer height variation which is less steep than that due to the adjacent updraft previously described. This mixed-layer height variation can be observed along the lagging edge of the mixed layer height field from 31 to 33 km east of Ocracoke in Fig. 6.

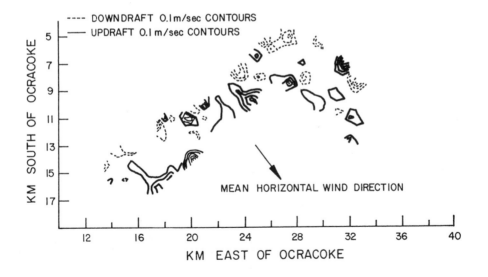

Figure 7.   Radar derived vertical wind contours at 300 meters ASL at 1456 GMT on March 2, 1986. The dashed contours are 0.1 m/sec downdrafts. The solid contours are 0.1 m/sec updrafts. for example, the core updraft, velocity 20 kilometers east of Ocracoke and 11 kilometers south of Ocracoke is 0.3 m/sec.

Radar-derived convergence contours in the horizontal plane at 100 m ASL are given in Fig. 8. Convergence as high as .003/sec was estimated along the leading edge of the chaff cloud. The mesa structures in the mixed-layer height field in Fig. 6 along the leading edge of the chaff cloud from 18 to 22 km and 30 to 33 km east of Ocracoke are coincident with the area of convergence in Fig. 8. The leading edge with a 0.002 m/sec convergence zone 22.5 km east of Ocracoke is associated with the 700 m ASL trough described by the mixed-layer height topography of Fig. 6.

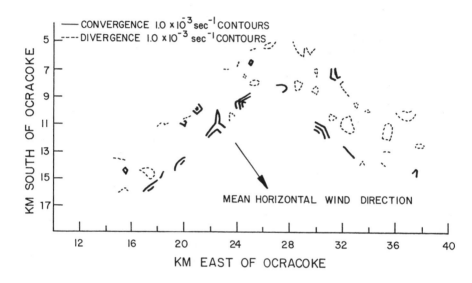

Figure 8. Radar derived divergence contours at 100 meters ASL at 1456 GMT on March 2, 1986. The dashed contours are 0.001 sec^{-1} divergence. The solid contours are 0.001 sec^{-1} convergence. For example, the core convergence 21 kilometers east of Ocracoke and 10 kilometers south of Ocracoke is 0.002 sec^{-1}.

The radar-derived eddy flow field of the 1456 GMT scan at 100 m ASL is shown in Fig. 9. The eddy field is calculated by subtracting the mean horizontal wind over the 0.5 km horizontal wind field grid at 100 m ASL from the wind vector at each grid point. The wide area of outflow associated with the lagging edge of the chaff cloud apex is striking. The areas of leading edge inflow are associated with the convergence zones shown in Fig. 8. The area of lagging edge outflow 37 km east of Ocracoke appears to have horizontally stretched the chaff cloud to the northeast. If the leading edge mixed-layer height field at 36 km east of Ocracoke in Fig. 6 is compared to the eddy field in Fig. 9, the two figures seem to indicate a mesa structure being formed in the mixed layer height field.

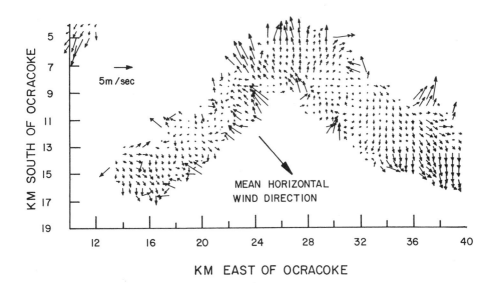

Figure 9.   Radar derived eddy flow field at 100 meters ASL at 1456 GMT on March 2,
            1986.

Fig. 10 is a plot of vertical wind contours in a vertical plane 6.0 km south of
Ocracoke. The plane is in the west-east direction. The dashed lines are the downdrafts
in 0.1 m/sec increments. The updrafts are represented by the narrow solid lines in
0.1 m/sec increments. The bold solid lines are the 0.0 m/sec contours. As the reader
views Fig. 10 from left to right, a west to east plane through the three dimensional
vertical wind field is observed. Fig. 10 describes the vertical wind structure along the
lagging edge of the mixed layer height field as described 6 km south of Ocracoke in Fig.
6. Notice in Fig. 10 the three 1.5 km wide areas of downdraft separated by the two
narrow weak updrafts. The downdraft core velocities are as high as 0.5 m/sec. The
updraft velocities are at the most an order of magnitude below these downdraft
magnitudes. The weak updraft region between 27.0 and 27.5 km east of Ocracoke is
the north west edge of an updraft region supporting the ridge in the mixed layer height
field between 7 and 8 km south of Ocracoke.

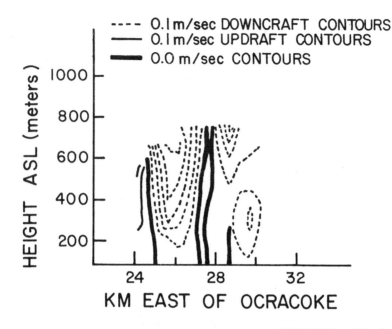

Figure 10. Radar derived vertical wind contours at at 1456 GMT on March 2, 1986.
This is in the vertical plane 6 kilometers south of Ocracoke. The dashed
lines are 0.1 m/sec downdraft contours. The bold solid lines are the
0.0 m/sec vertical wind contours. The thin solid lines are 0.1 m/sec updraft
contours. For example, there is a 0.5 m/sec core downdraft at 600 meters
ASL 26 kilometers east of Ocracoke.

4. CONCLUSIONS

The mesoscale structure of the marine boundary layer can be described in terms of
three-dimensional velocity and spatial structure as measured by two radars observing the
same chaff cloud.

The effect of convective mixing appeared to reduce baroclinic effects in the mixed
layer. A vertical plot of the mean horizontal wind averaged over 0.5 km radar derived
wind grid every 100 m ASL for the entire 1456 GMT chaff cloud indicates the layer to
be well mixed. The windspeed averaged 6.0 m/sec (+/-0.1 m/sec) up to 900 m ASL.
The wind direction averaged 292 deg (+/-1.0 deg) throughout this same layer. Well-
organized roll vortices were not observed. In fact the mesoscale mixed layer height
variations were in the mean wind direction. These variations in the height of the mixed
layer would be found in the cross wind direction if the secondary roll vortice circulation
had been present.

The spatial wavelengths of the entrainment layer processes appear to be smaller
than the resolution of the three dimensional cartesian wind field derived from the dual
radar data. These dimensions were 100 m in the vertical and 500 m in the horizontal. A
shorter radar baseline would have improved the spatial resolution of the radar measurements in the entrainment layer. This experimental design trade off would also results in
a dual doppler coverage area of smaller horizontal extent.

ACKNOWLEDGMENT

This work was supported by the National Science Foundation under the grant ATM-83-11812.

REFERENCES

Arya, S. P. S. and Wyngaard, J. C., 1975: Effect of Baroclinicity on Wind Profiles and the Geostrophic Drag Law for the Convective Planetary Boundary Layer. Journal of the Atmospheric Sciences, 32, 767-778.

Boers, R.: personal communication. NASA/GLAS/GSFC, Greenbelt, Maryland.

Brown, R. A., 1970: A Secondary Flow Model for the Planetary Boundary Layer. Journal of the Atmospheric Sciences, 27, 742-757.

Dirks, R. A., Kuettner, J. P., and Moore, J. A., 1988: Genisis of Atlantic Lows Experiment (GALE): An Overview. Bulletin of the American Meteorological Society, 69(2), 148-160.

GALE, 1985: Experiment Design Document, Genisis of Atlantic Lows Experiment, GALE Project Office, Boulder, Colorado, p. 79.

Huschke, R. E. Glossary of Meteorology, AMS, 1980, p. 174.

Kropfli, R. A., 1983: A Review of Microwave Radar Observations in the Dry Convective Planetary Boundary Layer. Boundary Layer Meteorology., 26, 51-67.

Kropfli, R. A. and J. M. Wilczak, 1986: Characteristics of Terrain Forced Eddies in the Marine Planetary Boundary Layer, From Doppler Radar Observations and a Mesoscale Numerical Model. Proceedings from the 23rd Radar Meteorology Conference, American Meteorological Society, JP186-JP189.

LeMone, M. A., 1973: The Structure and Dynamics of Horizontal Roll Vortices in the Planetary Boundary Layer. Journal of the Atmospheric Sciences, 30, 1077-1091.

Raman, Sethu and Riordan, Allen J., 1988: The Genisis of Atlantic Lows Experiment: The Planetary Boundary Layer Subprogram of GALE. Bulletin of the American Meteorological Society, 69(2), 161-172.

Schlesinger, R. J. 1961: Principles of Electronic Warfare (New York; Prentice Hall), 213 pp.

STUDY OF TALL CONVECTIVE STORMS
IN NORTH INDIA DURING SUMMER

R.N. Chatterjee, P. Prakash and R.K. Kapoor
Indian Institute of Tropical Meteorology
Pune–411005, India

ABSTRACT

Thunderstorms with great vertical extent have been observed in northern India at times. Frequency of such storms is maximum during summer. This paper describes some aspects of 161 thunderstorms whose radar echo tops reached or extended beyond 12 km and which were observed within 100 km around Delhi during the summer seasons of the 20-year period from 1959 to 1978. Based on the echo structure and echo intensity, these tall storms have been presumed to be of severe type. A study of storm height distribution showed that, in about 52 per cent of the storms, their tops lay between 12 and 13 km and, in 7.9 per cent of the storms, their tops penetrated the tropopause. The maximum height of the storm tops observed was 20 km. Their spatial distribution by octants showed that there is no preferential region of severe storm development around Delhi. It was also observed that severe thunderstorms have a tendency to organize themselves in lines. A comparative study of the composite soundings of two classes of days namely, days on which these tall storms occurred and days on which maximum height of the cloud tops were considerably lower (6-10 km) showed no significant difference in the mean temperature soundings between the two classes of days. However, mean mixing ratio was found to be significantly higher at all levels on days when tall storms occurred. Details of the study will be discussed.

1. INTRODUCTION

It is well known that thunderstorms can grow to great vertical extents, at times peeping into the stratosphere. These large thunderstorms may occur either individually or more typically in groups. Extensive damage to life and property, as well as the aviation hazards caused by the severe weather phenomena associated with these storms, needs no exaggeration.

Studies on the occurrence and mechanisms of thunderstorms have been a scientific goal for over three decades. The first extensive studies in this field were reported by Byers and Braham as early as 1949. Since then a number of studies have been reported (e.g., Battan, 1980; Brandes, 1977; Chatterjee and Prem Prakash, 1986; Cynthia and Carbone, 1987; Henry, 1964; King, 1980; Miller and Betts, 1977). In this paper an attempt has been made to study only those thunderstorms whose radar echo tops reached or exceeded 12 km, using the radar observations made within 100 km around Delhi (northern India) during the summer seasons (April - June) of the 20-year period from 1959 to 1978. Based on some of their observed radar echo features, such as sharply defined edges, well developed vertical structure, and high echo intensity, these tall storms have been considered severe in nature. The purpose of this study is to examine some general characteristics of severe thunderstorms in this region namely, height and spatial distribution, echo configuration, and tropopause penetration and to investigate atmospheric thermodynamic conditions responsible for the development of such storms. Details of the study are presented.

2. EQUIPMENT USED AND DATA COLLECTED

An X-band high-power Japanese radar of type NMD-451A was used. The characteristics of the radar set are given in Table 1. The radar was operated mainly from 1000 to 1700 IST (0430 to 1130 GMT). On each occasion, PPI scan at low elevation angles (near 0°) was first made to survey the precipitation occurrences and their general features around Delhi. The echoes were then picked up at random and their heights were measured on RHI or REI. From the radar observations made during the summer seasons of the 20-year period from 1959 to 1978, the data of convective type echoes within 100 km of Delhi (28° 37'N, 77° 12'E, 217 m MSL) were separated. Those convective echoes that attained maximum height of 12 km or more, and categorized as severe storms, have been considered in this paper. In all, there were 161 such storms, which constitute only 10.1 per cent of all the convective clouds (1,601) sampled during the above period.

TABLE 1.   RADAR CHARACTERISTICS

Wave length	3.2 cm
Peak power transmitted	250 KW
Pulse length	1 μ sec
Minimum detectable signal	-90 dBm
Pulse repetition frequency	300 Kz
Horizontal and vertical beam width	1.2°

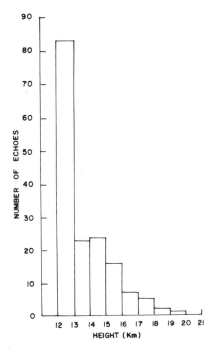

Figure 1.   Distribution of maximum echo
top heights of 161 storms.

3.   RESULTS AND DISCUSSIONS

3.1   HEIGHT DISTRIBUTION

The distribution of maximum echo-top heights of the 161 storms is shown in Fig. 1.   It may be seen from the figure that height of the echo tops of as many as 83 storms (51.6 per cent) lay between 12 and 13 km.   No echo reached a height greater than 20 km, and only 15 storms (9.3 per cent) grew beyond 16 km (average tropopause level of this region).

3.2   SPATIAL DISTRIBUTION

To assess whether there is any preferential region for the development of severe storms within 100 km of Delhi during the summer season, the region was divided into octants.   The frequency distribution of storm echoes in various octants is shown in Fig. 2. Considering storm echo frequencies in different octants, application of Chi-square test showed that the echoes were uniformly distributed

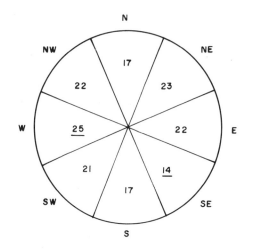

Figure 2.   Spatial distribution of 161 storms.

over the region.   Comparatively high concentration to the west (25)
and low concentration to the southeast (14) were significant only at
25 percent and 15 percent level, respectively.   The uniformly
distributed frequency pattern of the storm echoes during the summer
season stems from the fact that, in this season, the convective
activity is due to insolation and the thunderstorms are of the air
mass type that occurs randomly over the area.   Also, the region
considered is not influenced by any orographic effect.

3.3   DISTRIBUTION OF STORM ECHO TOPS RELATIVE TO THE TROPOPAUSE

     Effect of the tropopause to the vertical development of these
storms was investigated.   For this purpose, tropopause heights of
all the days when severe thunderstorms occurred have been considered
individually.   Tropopause heights were determined from the upper air
ascents at Delhi.   Upper air data for both 0530 IST (0000 GMT) and
1730 IST (1200 GMT) ascents were considered, and the height of the
tropopause on any day has been taken as the height obtained either
from 0530 or 1730 IST ascent, whichever was higher.   The days on
which the tropopause data of only single ascent were available, the
tropopause heights were determined by considering those single
ascents.   Days on which tropopause data were not available were not
considered.   For this reason, 10 of 161 storms were excluded, and the
remaining storms have been considered for the study.   The
distribution of the heights of the echo tops relative to the
tropopause of the 151 storms is shown in Fig. 3.   It can be seen from
the figure that only 12 storms (7.9 per cent) penetrated the actual
tropopause and there is a marked decrease in the frequency with
height above the tropopause.   The feature noticed suggests that
the tropopause could act as a barrier to the vertical development

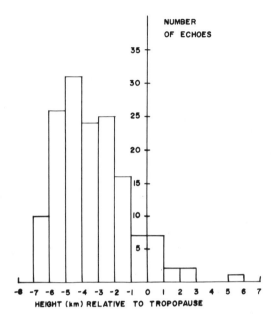

Figure 3. Distribution of maximum echo-top heights
           of 151 storms relative to the tropopause.

of the storms.   These observations seem to be in general agreement
with those of Henry (1964) in Alberta.   Maximum penetration observed
was 5.3 km.

3.4 HEIGHTS OF STORM ECHO TOPS VIS-A-VIS THEIR ECHO CONFIGURATION

        An investigation has also been made regarding the vertical
development of storms in relation to their echo configuration.   For
this purpose, storm echoes have been classified into different
height groups at 2-km intervals.   Then, in each height group, they
were further classified into three categories, namely (1) isolated,
(2) cluster and (3) line.   Storm echoes of the first category
consisted of those that were completely isolated from others, i.e.,
no other radar echo being present within 30 km.   Echoes of the second
category were associated with one or more neighboring radar echoes
within a distance of 30 km.   Echoes of the third category were part
of line of echoes consisting of three or more echoes.   Percentage
frequency distribution of storm echoes among different categories in
each height group is given in Table 2.   It can be seen from the table
that out of 161 storms, only 9 percent were of isolated type.   The

TABLE 2. PERCENTAGE FREQUENCY DISTRIBUTION OF SEVERE STORMS OF
DIFFERENT ECHO-TOP HEIGHTS AMONG DIFFERENT ECHO CATEGORIES

Height of echo top (km)	Category			No. of cases
	Isolated	Cluster	Line	
12.0 - 14.0	8	72	20	106
14.1 - 16.0	10	50	40	40
16.1 - 18.0	17	25	58	12
18.1 - 20.0	33	0	67	3
Total	9	62	29	161

rest of the storms were either cluster or line type (62 percent and
29 percent, respectively). That is, 91 percent of 161 storms were
embedded in other neighboring echoes and, there- fore, they grew in
the midst of other echoes. Thus, their immediate environment was
apparently very moist. This suggests that environmental moisture
condition plays a major role in the development   of severe
thunderstorms. It can be seen from the table that percentage of
storms of line type increased progress- ively with the height of the
storm echo tops. Simultaneous decrease in the occurrence of cluster-
type echoes (column 2) with height is noticed. This feature suggests
that very tall thunder-storm cells have a tendency to organize
themselves in lines.

3.5 ATMOSPHERIC THERMODYNAMICAL CONDITIONS

     In order to determine whether there existed any distinctive
features in the atmospheric thermodynamic conditions on days on
which severe thunderstorms occurred, a comparative study of
temperature and moisture distribution with height has been made for
the two classes of days namely, days on which storms with echo tops
of 12 km or higher occurred and days on which heights of the echo
tops were considerably lower (6-10 km). For this purpose, radar and
upper air data of Delhi (0530 IST) of only the 5-year period from
1966 to 1970 have been considered.   There were 15 days when the
heights of the echo tops reached or exceeded 12 km, and 19 days on
which heights of the echo tops lay between 6-10 km.   Mean vertical
profiles of temperature and mixing ratio, prepared for the above two
classes of days, are shown in Fig. 4. The figure shows that the
temperature, in general, was slightly higher from surface to about 14
km on days when tall storms occurred. A similar trend has also been
observed in the mean mixing ratio values at different levels. Also,
the difference in the mean mixing ratio values at different levels
between the two classes of days were more marked, compared with the
differences in temperature.

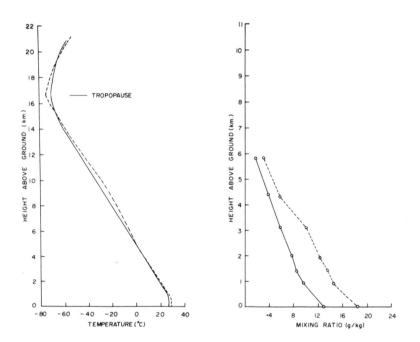

Figure 4. Profiles of mean temperature and mixing ratio
           for days on which echo heights lay between 6
           and 10 km (solid lines) and for days on which
           echo heights were 12 km or more (broken lines).

Further, two stability indexes namely, Showalter and K also
have been evaluated for each day, and the mean value of each index
with its standard deviation, for the two classes of days is given in
Table 3.   The table shows that the mean value of Showalter index was
more negative and the mean value of K-index was significantly higher
on days associated with tall storms.   The above result shows that
both Showalter index and K-index are good indicators for forecasting
severe storm development in this region during summer season.

4.   CONCLUSION

A radar study on the occurrence of severe thunderstorms in
northern India during summer season revealed the following:

(i)   About 10 per cent of the convective clouds forming
        around Delhi during summer develop into severe
        thunderstorms.

(ii)   More than half of the storms occurring in this region
        mature and dissipate below 13 km.   About 8 per cent
        penetrate tropopause.

TABLE 3. STABILITY INDEXES COMPUTED FROM THE MORNING (0000 GMT)
UPPER AIR ASCENTS AT DELHI FOR THE TWO CATEGORIES OF DAYS
(Figures within brackets indicate number of days)

	Showalter Index		K - Index	
	Days when heights lay between 6-10 km	Days when heights reached or exceeded 12 km	Days when heights lay between 6-10 km	Day when heights reached or exceeded 12 km
Mean	-0.5(19)	-2.7(15)	29(19)	40(15)
Standard Deviation	1.8	3.1	6.0	1.0

(iii)   Maximum height of the storm echo observed was 20 km.

(iv)   Spatial distribution of severe storms in Delhi region
         showed no preferential region for such storm
         development.

(v)   Very tall thunderstorms organize themselves in lines.

(vi)   Study of the mean vertical profiles of temperature
         and mixing ratio for the two classes of days namely,
         days on which storms of height 12 km and higher
         occurred and days on which echo tops were
         considerably lower (6-10 km) showed higher values of
         temperature as well as mixing ratio at all levels on
         days associated with tall storms, compared with days
         on which heights of the storm tops were comparatively
         lower.

ACKNOWLEDGMENT

The authors gratefully acknowledge the help received from
Messrs. R. B. Bhandari and S. P. Singh in preparing the paper.

REFERENCES

Battan, L. J., 1980:   Observations of two Colorado thunderstorms by
    means of zenith-pointing Doppler radar.   J. Appl. Meteor.,
    19, 580-592.

Brandes, E. A., 1977:   Flow in severe thunderstorms observed by dual
    Doppler radar. Mon. Wea. Rev., 105, 113-120.

Byers, H. R., and R. R. Braham, 1949:   Thunderstorms, U.S.
    Government Printing Office, Washington, D.C. 287 pp.

Chatterjee, R. N., and Prem Prakash, 1986:   A radar study on the
    frequency of occurrence of cumulonimbus clouds around Delhi.
    Mausam, 37, 241-244.

Cynthia, K. M., and E. Carbone, 1987:   Dynamics of thunderstorm
    outflow. J. Atmos. Sci., 44, 1879-1898.

Henry, C. D., 1964:   High radar echoes from Alberta thunderstorms.
    Scientific Report MW-38, MC Gill University, 105-126.

King, R. H., 1980:   Some characteristics of thunderstorm radar
    echoes in Northern Finland.   Preprints, 19th Conf. Radar
    Meteor. Miami, Amer. Met. Soc., 425-430.

Miller, M. J., and A. K. Betts, 1977:   Travelling convective storms
    over Venezuela, Mon. Wea. Rev., 105, 833-848.

SUBJECT INDEX*

*Number refers to the first page of the appropriate paper.